# A Year Without a Winter

# A Year Without a Winter

**Dehlia Hannah**
Editor

**Brenda Cooper**
**Joey Eschrich**
**Cynthia Selin**
Fiction Editors

## Contents

Fiction

This image is a cosmographic for *A Year Without a Winter*. Read counterclockwise beginning with the cherub blowing down a cloud of air, each figure signifies a calendar month. At February, a groundhog anticipates the end or continuation of winter. March is a "grue" emerald, which stands for the philosophical problem of induction, asking why past experiences justify future expectations. At April the element of water appears as a great wave moving across the horizon. An antique barometer indicates changes in atmospheric pressure at May. June is the month of the palm tree, a conflicted image of paradise. At July the earth element erupts into the sky, sending up clouds from below to mix with the winds blowing down along the vertical axis. August is Friedrich Schlegel's hedgehog, a philosophical animal in which fragment and totality are reconciled through art. From September a black crow looks back to the gemstone, David Hume's old riddle of induction giving way to Nelson Goodman's new one. Fire appears at October emanating from a face in the sun, or perhaps the moon, which exerts its attraction upon the rising tide across the horizontal. At November a satellite tracks weather patterns and facilitates instantaneous communication across the axis of technology. Mysterious hominid bones at December evoke Alfred Russel Wallace's search for the missing link to imagined hominid ancestors on the Malay Peninsula, where he happened to visit Tambora. As this cursory reading indicates, relations between these figures may be read through the multiple geometries traced by the atmosphere's meridian lines.

## Introduction

Dehlia Hannah

> "Prepare to hear of occurrences which are usually deemed marvelous. Were we among the tamer scenes of nature I might fear to encounter your unbelief, perhaps your ridicule; but many things will appear possible in these wild and mysterious regions which would provoke the laughter of those unacquainted with the ever-varied powers of nature."
>
> —Victor Frankenstein
> *Frankenstein: or, The Modern Prometheus*

Spoken aboard a ship in the Arctic Ocean, so begins the story of a dangerous force unleashed upon the world by human agency. Victor Frankenstein's words double as a prelude to another *unthinkable* tale: today, one hears of freak storms, dead zones, stilled winds, and record heatwaves not only at the world's geographical extremes but in its historically temperate zones. That these occurrences may be attributed to a monster of our own making defies easy belief. But climate change is not a fiction. On the bicentennial of its publication, Mary Shelley's *Frankenstein: or, The Modern Prometheus* (1818) remains modernity's most enduring parable of the risks inherent in disturbing the order of nature. Written amid a mysterious climate crisis, remembered as the Year Without a Summer, the novel registers the unpredictable impact upon the literary imagination of an atmosphere in disarray. Just as the narrative itself continues to haunt our encounters with emerging science and technology, the conditions of its production offer an exemplary case study of creation from within a global environmental catastrophe of which awareness and comprehension was necessarily incomplete. It may seem inexplicable to future generations that, as of 2018, the masses have not yet overwhelmed the streets in protest against an intolerable future. In the face of ample information and dire predictions, it seems that we suffer a collective failure of imagination of the crisis unfolding around us.[1] Inspired by the historical precedent of *Frankenstein*, this book chronicles a collective thought experiment designed to deepen our acquaintance with the ever-varied powers of nature and consider what stories and images might be conceived during *A Year Without a Winter*.

*Frankenstein* was immediately met with wild success—reprinted, translated, and adapted to the stage. Asked how an eighteen-year-old woman could have possibly written such a horrifying tale, Shelley recounts, in the introduction to the 1831 edition of the text, a memorable story of her adventures on the Grand Tour in June 1816. Amid a scandalous romance, the young Mary Godwin and her married lover Percy Bysshe Shelley took up residence near Lord Byron on the banks of Lake Geneva, where the first stirrings of unease at the ominous weather were already palpable on the land. "At first, we spent our pleasant hours on the lake or wandering on its shores...," she writes, while Byron stoked their literary ambitions in salons and parties at the Villa Diodati. "But it proved a wet, ungenial summer, and incessant rain often confined us for days to the house. Some volumes of ghost stories, translated from the German into French, fell into our hands... 'We will each write a ghost story,' said Lord Byron; and his proposition was acceded to." This friendly challenge was remembered as the "dare" that spawned not only two great literary monsters but the new genre of science fiction. Inspired by violent thunderstorms, lectures on the latest scientific theories, spirits of laudanum, and the Gothic, Shelley was gripped by a reverie: she envisioned a hideous human form, assembled from inanimate parts and brought to life by artificial means, only to be

1 "Where are the books? The poems? The plays? The goddamn operas?" pleaded Bill McKibben in "Imagine That: What the Warming World Needs Now Is Art, Sweet Art," *Grist*, April 21, 2005. A decade later they are abundant; a deeper diagnosis is offered by Amitav Ghosh in *The Great Derangement: Climate Change and the Unthinkable* (Chicago: University of Chicago Press, 2016).

**Engraving of the Villa Diodati, by Edward Finden after William Purser, 1832.**

abandoned in disgust and regret by his creator. *The Vampyre*, the story of a bloodthirsty nobleman and a thinly disguised portrait of Byron, was penned by his doctor, John Polidori. Their own attempts to write ghost stories abandoned, Lord Byron and Percy Shelley's writings from the summer of 1816 betray a sense of anxious attention to the environment looming at the fine edge of sublimity and fear. Byron's poem "Darkness" depicts a world in which birds take to roost at midday under the dim embers of a dying sun, as the earth gradually loses its heat to a coming Ice Age. Percy Shelley's *Mont Blanc: Lines Written in the Vale of Chamouni* offers an ode to the unfathomable powers of nature embodied in a mountain.

Shelley and her companions could hardly have imagined that the incommodious weather they endured on their early summer holiday was not merely a local aberration. But as unseasonal frosts and precipitation swept across Europe, crops withered in the fields and food became scarce. An atmosphere of terror took hold as news spread of similar conditions afflicting Russia and North America, where snow fell in July along the Eastern Seaboard. Conjecture about the cause of the apocalyptic weather was matched by equally wild speculation in the marketplace, causing social upheaval and compounding an emerging humanitarian crisis. No one knew if or when the Year Without a Summer would end. In fact, it lasted for three more years before seasonal cycles began to approach normality in 1818, just as *Frankenstein* went to press—finding a receptive readership, perhaps, in a traumatized public. The truly global extent of the catastrophe would only become clear in retrospect, however, once the cause of the long summerless "year" was definitively established. As David Higgins argues in this volume, an assemblage of disparate voices from that time can now be gathered into a coherent narrative of the crisis, as registered from within its midst.[2]

A volcano had long been suspected. But it took the eruption of Mount Pinatubo in 1991, and the observation of similar climatic effects, for a new global narrative to coalesce from available historical and geophysical records, casting the novel's origins in a new light and amplifying the significance of its environmental themes.[3] We now know that the summer of 1816 was the beginning of a three-year period of global cooling caused by the eruption, in April 1815, of Mount Tambora, on the Indonesian island of Sumbawa. The largest volcanic eruption ever recorded, Tambora caused vast destruction locally and disturbed weather patterns for years to come. Over the course of several days of explosions, its peak was reduced by over a thousand meters, sending an estimated fifty cubic kilometers of dense rock and superheated gases high into the sky. At first the sounds were attributed to canon fire, but visual clues soon offered an alternative explanation. Ashes fell like snow, and "the atmosphere forboded an Earthquake," according to the reports collected by the colonial governor and avid naturalist Sir Stamford Raffles, blanketing surrounding islands in a dark, unbreathable fog.[4] In its immediate aftermath, over seventy thousand people are estimated to have died on Tambora's slopes, an entire kingdom buried beneath mudslides of burning tephra. Hundreds of thousands more would find the

---

2 See also David Higgins, *British Romanticism, Climate Change, and the Anthropocene: Writing Tambora* (New York: Palgrave Macmillan, 2017).

3 Clive Oppenheimer, "Climatic, Environmental, and Human Consequences of the Largest Known Historic Eruption: Tambora Volcano (Indonesia) 1815," *Progress in Physical Geography: Earth and Environment,* vol. 27, no. 2 (June 2003): 230–259; Gillen D'Arcy Wood, *Tambora: The Eruption That Changed the World* (Princeton NJ: Princeton University Press, 2014).

4 Sir Thomas Stamford Raffles, *Narrative of the Effects of the Eruption from the Tomboro Mountain in the Islands of Sumbawa on the 11th and 12th of April 1815* (1816).

local land uninhabitable for decades to come. The eruption was powerful enough to eject dust and superheated gases into the upper atmosphere, where they were circulated around the globe by stratospheric winds. Within months, the monsoon that regulates seasonal rainfall patterns on the Indian subcontinent and in the Asia Pacific region was disrupted, bringing about catastrophic flooding, famine, and the spread of epidemic disease. By the following year, sun-blocking volcanic dust and sulfur dioxide had cooled the planet by an average of one degree Celsius, a temperature change whose regional effects were much more dramatic. Under Tambora-darkened skies, sunspots became visible to the naked eye for the first time in many regions of the world, and strange colors filtered through the air. Despite the cold temperature, powerful shifts in wind and ocean currents allowed the Arctic ice cap to fracture, tempting explorers in search of the fabled Northwest Passage toward the North Pole, where Shelley locates the opening scenes of *Frankenstein*.

Occurrences of the sort that fueled fantasies of environmental opportunity and collapse two centuries ago once again saturate our news media. Wildfires burning above the Arctic Circle, great sinkholes melting through the Siberian permafrost, airplane tires melting on the tarmac, cities counting down to "day zero" water supplies, whole island nations submerged by rising seas; these scenes offer sneak previews of the coming nightmare of global seasonal arrhythmia. The cause of the long-term warming trend that we are experiencing today is well known: in the intervening period since Tambora, humans have unearthed and burned vast quantities of coal and oil, cleared forests, and released enormous quantities of carbon dioxide into the atmosphere. There, it traps the sun's warmth like the glass walls of a greenhouse. Whereas those who lived in Tambora's shadow could not possibly know the scope of the catastrophe, not least because they lacked the means of communications to compare observations in real time, predictions of anthropogenic climate change have been made since 1896 and have steadily gained empirical support.[5] The science of meteorology was fundamentally transformed in

5 Paul N. Edwards, *A Vast Machine: Computer Models, Climate Data, and the Politics of Global Warming* (Cambridge, MA: MIT Press, 2010).

the twentieth century by the development of a global weather monitoring and communications infrastructure, modern statistical methods and computing power, and ultimately by the emergence of Global Circulation Models (GSMs) of the dynamical behavior of the earth's atmosphere. Only within this context could the abstract concept of *global* climate emerge, one that would allow the notoriously variable phenomenon of local weather to be situated within planetary scale systems of wind patterns, ocean circulation currents, and atmospheric chemistry. The greenhouse effect has already raised the global mean temperature by one degree Celsius. As stated in the nonbinding Paris Agreement of 2015, the most ambitious goal of the international community is to limit warming to two degrees. This goal may be highly unrealistic given that almost all of the Intergovernmental Panel on Climate Change (IPCC) scenarios upon which this agreement is based presuppose "negative emissions," not only reducing emissions but removing existing CO2 from the air, the technological means for doing which does not yet exist.[6] Envisioning a need for stop-gap measures to mitigate the worst effects of climate change, engineers entertain perhaps Promethean ambitions to cool our rapidly warming planet by intentionally mimicking the climatic effects of volcanic eruptions through Solar Radiation Management (SRM). Our best climate models carry high levels of uncertainty, but one thing is clear: ours will be a future without familiar seasons.

In the context of a slowly dawning awareness concerning the complex implications of anthropogenic climate change, Gillen D'Arcy Wood's compelling history, *Tambora: The Eruption that Changed the World*, offers an urgent reminder of the relevance of that historical moment to our own:

> On the eve of Tambora's bicentenary, and facing multiplying extreme weather crises of our own, its eruption looms as the richest case study we have for understanding how abrupt shifts in climate affect human societies on global scales and decadal time frames. The Tambora climate emergency of 1815–1818 offers us a rare, clear window onto a world

6 Abby Rabinowitz and Amanda Simson, "The Dirty Secret of the World's Plan to Avert Climate Disaster," *Wired*, December 10, 2017.

convulsed by weather extremes, with human communities everywhere struggling to adapt to sudden radical shifts in temperatures and rainfall, and a flow-on tsunami of famine, disease, dislocation, and unrest.[7]

Put another way, the Tambora period can be framed as a natural experiment, a global-scale perturbation of the steady states of human and natural systems that affords scholars opportunities to identify causal relationships among variables and make predictions about the behavior of similar systems under similar conditions.[8] Using such methods of comparison, volcanic eruptions are often treated as "natural experiments" by historians and climatologists, whose studies offer both the basis for and serious arguments against proposed geoengineering field experiments.[9]

If climate change eludes our imaginative grasp, it is perhaps because, in contrast to dramatic events like earthquakes and eruptions, its effects unfold in the atmosphere, hydrosphere, biosphere, and geosphere over a multiplicity of time scales and processes, some of which are beyond the scope of human sensory perception or even life span.[10] *A Year Without a Winter* uses the bicentenary of Tambora's three-year climate crisis as the basis for delimiting 2015–2018 as a sample of our ongoing climate emergency. Constraining an indefinite process within a more manageable time frame, its agenda pursues point-by-point comparisons, allowing similarities and dis-analogies to become identifiable.

---

**7** Wood, *Tambora*, 230.

**8** Jared Diamond and James A. Robinson, eds., *Natural Experiments of History* (Cambridge, MA: Belknap Press of Harvard University Press, 2010).

**9** See, for example, David W. Keith, Riley Duren, and Douglas G. MacMartin, "Field Experiments on Solar Geoengineering: Report of a Workshop Exploring a Representative Research Portfolio," *Philosophical Transactions of the Royal Society of Mathematical, Physical, and Engineering Sciences*, vol. 372, no. 2,031 (December 28, 2014); Jack Stilgoe, *Experiment Earth: Responsible Innovation in Geoengineering* (New York: Routledge, 2014).

**10** On the incommensurate temporalities of environmental catastrophes, see Dipesh Chakrabarty, "The Climate of History: Four Theses," *Critical Inquiry*, vol. 35, no. 2 (Winter 2009): 197–222; Rob Nixon, *Slow Violence and the Environmentalism of the Poor* (Cambridge, MA: Harvard University Press, 2011); Robert Markley, "Time," in *Telemorphosis: Theory in the Era of Climate Change*, vol. 1, ed. Tom Cohen (Ann Arbor, MI: Open Humanities Press, 2012).

Considering overlaps between the Tambora period and the present foregrounds modes of attention that emerge within environmental crises and their literary and artistic expression, a study that serves as the basis for a field experiment of our own.

An impromptu writing retreat, during which four authors found inspiration under the unwelcome constraints imposed by a disordered environment, became a recipe for our own purposeful immersion in an unfamiliar landscape. For the bicentenary of the "dare," four science-fiction authors—Tobias S. Buckell, Nancy Kress, Nnedi Okorafor, and Vandana Singh—were commissioned to participate in a reenactment of the founding moment of the genre, with the aim of updating its parameters to reflect our contemporary environmental conditions and concerns. As in all historical reenactments, much was different in the 2016 iteration of the dare; what remained constant was the basic form of a sociable excursion to a shared residence. Ours lasted for two days, during which we explored our built and natural environs, shared meals and read aloud, and heard lectures on the latest scientific research. What differed most profoundly was that, whereas the Villa Diodati group sought to escape the foul weather of the Year Without a Summer, the curatorial logic of *A Year Without a Winter* placed the environment in the foreground.

Rather than returning to the Alps, a site was chosen in the Arizona desert, a place where issues of water scarcity, sustainable energy and land use loom large. At the same time, we sought to amplify our understanding of the global climate crisis unfolding around us by calling attention to the relations between our shared, local experience and distant locations linked by wind patterns, water resources, energy economies, and human and animal migration. In place of theories of Galvanism and glaciation, biogeochemist Hilairy Hartnett painted a haunting picture of the effects of climate change on the ocean, which responds much more slowly than the atmosphere to rising carbon concentrations. Based on a meteorological definition of "a year without a winter" as "the year when the coldest winters first become warmer than the warmest winters of the past," a preliminary analysis of IPCC scenarios conducted by Melissa Bukovsky of the National Center for Atmospheric Research suggested that such conditions could be anticipated in major European and American cities sometime around the third quarter of the twenty-first century. With these considerations in mind, futurist Cynthia Selin led us to immerse ourselves imaginatively in fictional scenarios of winterless futures.

A desert landscape offers fertile ground for envisioning a warmer world, one in which we will be reliant upon the built environment to mediate our exposure to a harsh climate. If, like Shelley, or the protagonist of Okorafor's story in this volume, one finds oneself confined to a house while the storm rages outside, then it matters very much just what sort of house it is. The experimental town of Arcosanti was designed by the architect Paolo Soleri in the 1960s, embodying a theory of urban planning that combined the insights of architecture and ecology in a megastructure with high population density and low environmental impact, or an *arcology*. Although Soleri's utopian aspirations have never been fully realized within the small constellation of buildings erected at Arcosanti since building began there in the 1970s, the image of an arcology has afforded rich inspiration to science fiction, cyberpunk, fantasy games, and designers of habitats for extreme environments from Antarctica to Mars. Excavated from Arcosanti's own archives by architectural historian James Graham, voices from the past tell of struggles to realize a just society within the newly imagined infrastructures. History and fiction alike suggest that the cultural affordances of our physical surroundings require endless renovation. Even on our short trip, group dynamics ebbed and flowed under the variable influences of food and wine, disturbed circadian rhythms and arachnophobia. Out of a cloudless sky, a thundershower appeared directly overhead just as we sat down to read "Darkness" under the stars. Returning from Arcosanti to Phoenix for a final dinner at the Desert Botanical Garden—tamer scenes of the landscape we had just visited—it was easy to imagine how minor inconveniences would have mounted in significance had the dare not come to a timely conclusion but continued indefinitely...

Other inversions and substitutions issued from the relocation of the "dare" to the desert. If the novel was the quintessential aesthetic form of the nineteenth century, in what form would the environmental imaginaries of the twenty-first century find their most compelling expression? Might new media or genres be born of our efforts to think through our hypermediated, modeled and predicted, yet highly uncertain climate futures? What other sites would yield telling stories and images of a "year without a winter"? Over the course of 2016, these questions were explored through an intensive series of conferences, publications, and exhibitions, designed to enroll a broad range of participants into a collective thought experiment about what "winter"—and other seasons—mean in various cultural contexts and what it would mean to lose them. "Unseasonal Fashion: A Manifesto" attended to trends in architecture and fashion design for reflections of tacit awareness of climate change, to outfits as quotidian expressions of compromise between enduring and desiring the weather.[11] In collaboration with curator Jacob Lillemose, a speculative scenario became the basis for the exhibition *Dressing in a World of Endless Rainfall*, which showcased garments designed by Anne van Galen to mediate an uncanny embrace of exposure to the elements—a condition easy to imagine in Copenhagen during the rainy summer of 2016. A conference on "Fictions and Policy" was convened at the European Commission's Joint Research Center in advance of the 22nd Conference of Parties (COP22) meeting in Marrakech, to examine how images and narratives shape the legislative landscape of climate change. On December 21, 2016, a public performance of Amy Balkin's *Reading of the Paris Agreement* was staged in Copenhagen to mark the winter solstice and reflect upon the divergence of astronomical and meteorological winter. For EMERGE, an annual festival of art and technology held in Phoenix, Karolina Sobecka was commissioned to develop *Cloud Services* (2017–present), a proposal to genetically modify bacteria that moderate ice-nucleation in rain clouds in order to allow their genomes to store digital data, enabling the hydrological cycle to operate as a passive data distribution system. With this modest yet technologically plausible proposal, Sobecka assumed the role of a modern Victor Frankenstein, becoming a flash point for public debate about the promises and perils of geoengineering. Asked to decode a message stored in a raindrop in Sobecka's deadpan mock documentary of *Cloud Services*, a scientist reads off a black screen, "You will rejoice to hear that no disaster has accompanied the commencement of an enterprise

---

**11** Dehlia Hannah and Cynthia Selin, "Unseasonal Fashion: A Manifesto," *Climates: Architecture and the Planetary Imaginary*, ed. James Graham et al. (Zurich: Lars Müller Publishers, 2016), 222–231.

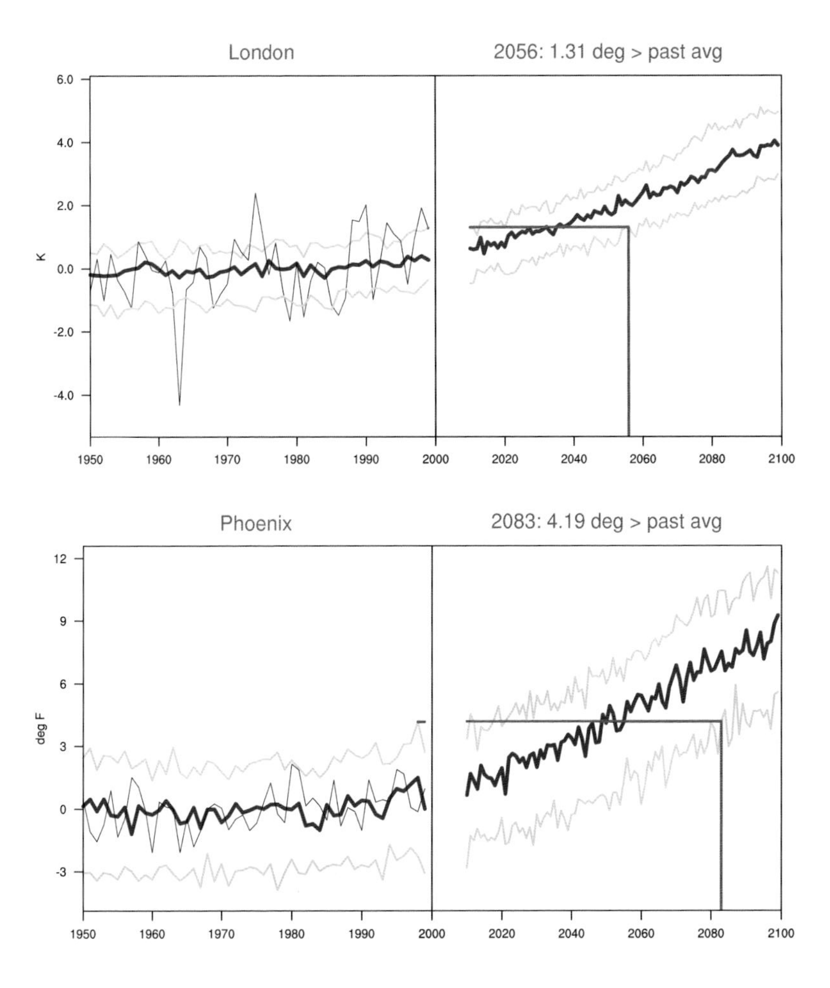


**Melissa Bukovsky, *A Year Without a* Meteorological *Winter*—the year when the coldest winters will become warmer than the warmest winters of the past. Dark gray lines represent the average temperature for a given year; the top blue line represents the warmest winters (75th percentile of the model ensemble) and bottom blue line represents the coldest winter (25th percentile of the model ensemble). Courtesy of Melissa Bukovsky.**

which you have regarded with such evil forebodings"—the first line of Shelley's novel, which flow from the pen of polar explorer Roger Walton.

Just as Victor pleads for the veracity of his account in the opening pages of *Frankenstein* by invoking his interlocutor's shared experiences of nature's extremes, belief in climate change is bolstered by visceral exposure to its effects.[12] But because it takes place on inhuman scales, isolated experiences, however powerful, must be organized within a coherent system of representation in order to belong to a unified picture of the world. As a representation of a vast dynamical system, a "picture" of climate and its changes requires a temporal dimension and reduction in scale, forms of simplification and abstraction that are achieved in earth systems sciences through models. In the arts, such representational imperatives are often enacted through narrative. Evoked through static images and objects pregnant with meaning, or told in durational media such as film and literature, narratives offer points of entry, identification, and extrapolation from a fictional space to a real world of similar characters and events. Indeed, philosopher Roman Frigg argues that we think about the world through scientific models in much the same way as we do through literary fiction, by imagining how characters or variables *would* behave within different possibility spaces, of which the real world is one.[13] While scientific models are constrained by data to produce realistic representations, novels have a greater degree of freedom, constrained only by the norms of genre. Amitav Ghosh has cogently argued that some of the blame for our failure of cultural imagination in the face of climate change falls to the realist novel, wherein the portrayal of implausible events constitutes a violation of literary convention sufficient to warrant expulsion to the lower ranks of science fiction. Realism cannot tolerate "the mysterious work of our own hands returning to haunt us in

---

**12** See, for example, Jane L. Risen and Clayton R. Critcher, "Visceral Fit: While in a Visceral State, Associated States of the World Seem More Likely," *Journal of Personality and Social Psychology*, vol. 100, no. 5 (May 2011); Stephen B. Broomell, David V. Budescu, and Han-Hui Por, "Personal Experience with Climate Change Predicts Intentions to Act," *Global Environmental Change* 32 (May 2015): 67–73.

**13** Roman Frigg, "Models and Fiction," *Synthese*, vol. 172, no. 2 (March 2009): 251–268.

unthinkable shapes and forms."[14] It seems, then, that the cultural sphere shares with economists a notorious failure to recognize in the real world phenomena that are not represented in the fictional space, respectively, of the novel or the financial model—leading, in both cases, to crisis.[15] In the case of climate change, this failure manifests as an incapacity to imagine and give aesthetic form to phenomena that *are* well represented in climate models. The consequence of our failure of imagination is not misunderstanding or denial of the facts but rather a difficulty situating ourselves within the untame and unnatural scenes of our changing global environs. We must return to the world itself to build the narrative again.

By early 2017, as the world entered its third year of record-breaking temperatures, a "year without a winter" began to seem less like a future scenario than a description of the present, as though the planet itself was reenacting the Tambora climate crisis in reverse. At this point, the project took on new geographical and creative trajectories. At the Armory Show in March 2017, I was captivated by the sight of a plant sheathed in ice inside a refrigerated glass vitrine, its walls encrusted with crystalline forms, as though a precious relic of winter had been preserved. Two weeks later, I found myself aboard a ship to Antarctica, where I envisioned myself retracing an inversion of Walton's fictive expedition to the North Pole. Within the framework of *A Year Without a Winter*, it could not be more fitting that the reason for my trip was to attend the 1st Antarctic Biennale, an art exhibition at a location that is exquisitely sensitive to climate change. Departing just days after a gigantic crack was detected in the Larsen C ice shelf, the Antarctic Biennale lends itself to narration as the proper reenactment of the "dare." There I chanced to meet the artist whose frozen plant now graces the cover of this book, Julian Charrière. What emerged from the conversations with artists, scholars, and the exhibition's curator, Nadim Samman, aboard that ship was a deeper understanding of how we inscribe ourselves into the world through image and narrative, eliding any stable distinction between the real and fictional; natural and artificial; past, present,

**14** Ghosh, *The Great Derangement*, 32.
**15** See Mary Morgan, *The World in the Model: How Economists Work and Think* (Cambridge, UK: Cambridge University Press, 2012).

and future. In Antarctica, struggling to erect temporary installations that would disappear without leaving a trace, amid whales, penguins, guano, and an international regulatory apparatus, it became evident that we are protagonists in a story as much as agents in the geophysical situation. Assuming responsibility for the dramaturgy of such stories, Samman suggests, is the twenty-first-century curator's project. From house to boat, mythic monsters to invasive species, the characters and scenography of *A Year Without a Winter*'s multiple science fictional worlds spilled forth into the real world. The fourth wall of the narrative shattered, the contributors to this book became actors on its stage, the planet a nonhuman participant in our collective thought experiment.

As fiction, history, and future scenarios blended seamlessly into narratives of the present, a final site asserted its prominence within the geographical imaginary of this project: Tambora. The volcano had figured as an important geophysical cause, but its cultural stories remained to be considered within the framework of this project. Of Sir Stamford Raffles's 1816 account of the eruption, assembled in its immediate aftermath as a report to the British colonial administration, Higgins observes that, "the narrative is not merely a commentary on a political-ecological catastrophe, but an intervention within it." Wood's 2014 account of Tambora's global social and cultural effects, assembled from sources ranging from archaeological investigations to Chinese poetry to epidemiological studies of the nineteenth-century cholera epidemic, likewise constitutes not merely a commentary on a past catastrophe but an intervention within our contemporary one. Included in this volume, Wood's personal reflections of his research trip to Tambora offer profound insights into how we inhabit landscapes through layers of literary, historical, and political narratives, narratives that our bodily presence also changes—both inadvertently and purposefully. The importance of fieldwork as an investigatory and a constructive practice runs a common thread through the methods of scientists, historians, curators, philosophers, writers, artists, political, and humanitarian workers who have participated in this project. These practices are not limited to experts at remote sites, however; the narratives we live by at home may be the most consequential. These became the subject of the COP23 workshop "Are We Listening? Rethinking

Narratives and Context for Climate Communication," led by Pablo Suarez of the Red Cross Red Crescent Climate Centre. In familiar landscapes as well as places that one encounters as already supersaturated with meaning, an openness for new stories to emerge must be carefully cultivated. With these considerations in mind, Charrière and I ventured to Tambora, where our attention was captured by another dark cloud whose global circulation through contemporary ecologies and economies is elemental to our contemporary climate crisis. Amid burning forests and vast plantations of oil palms, we learned that Tambora may be translated as "an invitation to disappear." Closing the narrative frame of Walton's letters, this book ends with our postcards from Tambora, an account of the beginning of a film and installation that borrows its name from the volcano.[16]

**16** Dehlia Hannah, "Field Philosophy," *Julian Charrière—An Invitation to Disappear* (Roma Publications, Musée des Bagnes & Kunsthalle Mainz, 2008).

# DARKNESS

Lord Byron

---

I had a dream, which was not all a dream.
The bright sun was extinguish'd, and the stars
Did wander darkling in the eternal space,
Rayless, and pathless, and the icy earth
Swung blind and blackening in the moonless air;
Morn came and went—and came, and brought no day,
And men forgot their passions in the dread
Of this their desolation; and all hearts
Were chill'd into a selfish prayer for light:
And they did live by watchfires—and the thrones,
The palaces of crowned kings—the huts,
The habitations of all things which dwell,
Were burnt for beacons; cities were consum'd,
And men were gather'd round their blazing homes
To look once more into each other's face;
Happy were those who dwelt within the eye
Of the volcanos, and their mountain-torch:
A fearful hope was all the world contain'd;
Forests were set on fire—but hour by hour
They fell and faded—and the crackling trunks
Extinguish'd with a crash—and all was black.
The brows of men by the despairing light
Wore an unearthly aspect, as by fits
The flashes fell upon them; some lay down
And hid their eyes and wept; and some did rest
Their chins upon their clenched hands, and smil'd;
And others hurried to and fro, and fed

Their funeral piles with fuel, and look'd up
With mad disquietude on the dull sky,
The pall of a past world; and then again
With curses cast them down upon the dust,
And gnash'd their teeth and howl'd: the wild birds shriek'd
And, terrified, did flutter on the ground,
And flap their useless wings; the wildest brutes
Came tame and tremulous; and vipers crawl'd
And twin'd themselves among the multitude,
Hissing, but stingless—they were slain for food.
And War, which for a moment was no more,
Did glut himself again: a meal was bought
With blood, and each sate sullenly apart
Gorging himself in gloom: no love was left;
All earth was but one thought—and that was death
Immediate and inglorious; and the pang
Of famine fed upon all entrails—men
Died, and their bones were tombless as their flesh;
The meagre by the meagre were devour'd,
Even dogs assail'd their masters, all save one,
And he was faithful to a corse, and kept
The birds and beasts and famish'd men at bay,
Till hunger clung them, or the dropping dead
Lur'd their lank jaws; himself sought out no food,
But with a piteous and perpetual moan,
And a quick desolate cry, licking the hand
Which answer'd not with a caress—he died.
The crowd was famish'd by degrees; but two
Of an enormous city did survive,
And they were enemies: they met beside
The dying embers of an altar-place
Where had been heap'd a mass of holy things
For an unholy usage; they rak'd up,
And shivering scrap'd with their cold skeleton hands
The feeble ashes, and their feeble breath
Blew for a little life, and made a flame
Which was a mockery; then they lifted up
Their eyes as it grew lighter, and beheld
Each other's aspects—saw, and shriek'd, and died—
Even of their mutual hideousness they died,

Unknowing who he was upon whose brow
Famine had written Fiend. The world was void,
The populous and the powerful was a lump,
Seasonless, herbless, treeless, manless, lifeless—
A lump of death—a chaos of hard clay.
The rivers, lakes and ocean all stood still,
And nothing stirr'd within their silent depths;
Ships sailorless lay rotting on the sea,
And their masts fell down piecemeal: as they dropp'd
They slept on the abyss without a surge—
The waves were dead; the tides were in their grave,
The moon, their mistress, had expir'd before;
The winds were wither'd in the stagnant air,
And the clouds perish'd; Darkness had no need
Of aid from them—She was the Universe.

# Tambora and British Romantic Writing

David Higgins

The connection between the eruption of Mount Tambora and the subsequent disruption to global weather patterns was not understood until the twentieth century, and, indeed, the eruption itself was not widely reported in Britain. Nonetheless, both processes had a significant impact on British Romantic writing. When the eruption took place in April 1815, Java was under the control of British forces, who had taken it from the Dutch in 1811, largely because of its strategic potential in the ongoing Anglo-French conflict. As a result, the principal source for eyewitness accounts of the eruption and its aftermath is an English-language document collected together under the auspices of Sir Stamford Raffles, the island's governor until 1816. In Britain and Europe, writers also produced some important works in response to the stormy and cold weather of the Year Without a Summer. Its traces have been identified in texts by authors such as John Keats, Eleanor Anne Porden, and particularly Lord Byron, Percy Bysshe Shelley, and Mary Shelley.[1] However, scholars of Tambora have tended to evoke these texts for illustrative rather than analytical purposes and therefore the full implications of the crisis for how we understand the

---

1 For Keats, see Jonathan Bate, *The Song of the Earth* (London: Picador, 2000), 102, and Alexandra Harris, *Weatherland* (London: Thames & Hudson, 2015), 258; for Porden, see Adeline Johns-Putra, "Historicizing the Networks of Ecology and Culture: Eleanor Anne Porden and Nineteenth-Century Climate Change," *Interdisciplinary Studies in Literature and Environment*, vol. 22, no. 1 (June 2015): 27–46; for Byron and the Shelleys, see Bate, *The Song of the Earth*, chapter 4, and Gillen D'Arcy Wood, *Tambora: The Eruption that Changed the World* (Princeton, NJ: Princeton University Press, 2014), especially chapter 3. Any scholar working on the cultural effects of Tambora is indebted to Wood's study.

relationship between humans and nonhuman nature have not been addressed. If Romantic writers were often excited by the human capacity to shape the world, they were also troubled by our vulnerability to elemental forces apparently beyond our control. They therefore speak to the increasingly influential idea that we are entering a new geological epoch characterized by the complex entanglement of human activity and earth systems: the Anthropocene.

Of course, this idea is not without controversy. Some scholars have criticized it for suggesting a species-wide agency that obscures the inequalities that have driven, and are driven by, global environmental change.[2] Others have suggested that it implies a dangerous anthropocentrism.[3] Its use by non-scientists has been challenged as detached from the specifics of the term as a stratigraphic marker that indicates a step change in Earth systems.[4] The start date has also been debated, with suggestions going as far back as the climatic impacts of agriculture ten thousand years ago and as far forward as the traces in the rock record left by nuclear testing after World War II.[5] Indeed, some geoscientists have identified the emergence of the Anthropocene in the Romantic period, noting the stratigraphic potential of the coincidence of the Tambora eruption and the early stages of industrial capitalism.[6] I see the Tambora crisis, when placed within the genealogy of the Anthropocene, not only in parallel to

---

**2** See, for example, Andreas Malm and Alf Hornborg, "The Geology of Mankind? A Critique of the Anthropocene Narrative," the *Anthropocene Review*, vol. 1, no. 1 (January 2014): 62–69.

**3** Stacy Alaimo, "Elemental Love in the Anthropocene," in Jeffrey Jerome Cohen and Lowell Duckert, eds., *Elemental Ecocriticism: Thinking with Earth, Air, Water, and Fire* (Minneapolis: University of Minnesota Press, 2015), 298–309.

**4** See, for example, Clive Hamilton, "The Anthropocene as Rupture," the *Anthropocene Review*, vol. 3, no. 2 (February 2016): 93–106.

**5** The latter date of around 1950 seems the most likely one to be taken up by the International Commission on Stratigraphy.

**6** See, for example, Jan Zalasiewicz, Mark Williams, et al., "Are We Now Living in the Anthropocene?" *GSA Today*, vol. 18, no. 2 (February 2008): 4–8, and Victoria C. Smith, "Volcanic Markers for Dating the Onset of the Anthropocene," in *A Stratigraphical Basis for the Anthropocene*, ed. C. N. Waters, J. A. Zalasiewicz, M. Williams, M. A. Ellis, and A. M. Snelling (London: Geological Society Special Publications, 2013), 283–299. For an overview of the various different starting dates that have been suggested, see Bruce D. Smith and Melinda A. Zeder, "The Onset of the Anthropocene," *Anthropocene*, vol. 4 (December 2013): 8–13.

our contemporary crisis but in connection to it. This essay also understands so-called natural disasters as inflected by politics, culture, and narrative. New materialism is an attractive body of work for such an approach because of its emphasis on relational thinking and the interactions (or, following Karen Barad, "intra-actions") of matter and meaning.[7] I have therefore tried to adopt a new materialist methodology appropriate to the Tambora crisis as a complex assemblage of human and nonhuman actors, by focusing not on individual authors but on reconstructing textual assemblages that can be understood retrospectively not merely as representing the crisis but intervening within it. For reasons of space, I focus here on two key texts: the narrative of the eruption compiled by the British administration in Java and Mary Shelley's novel *Frankenstein* (1818).[8]

### The Colonial Narrative

Scholars have tended to treat the Anglophone account of the eruption, first published in the *Transactions of the Batavian Society* in 1816, as a straightforward source of information.[9] However, it is in fact a heteroglossic and collaborative production in which different forms of knowledge intertwine. Sir Stamford Raffles's desire to mediate and control representations of the eruption within a single centralizing document was connected to his role as a colonial administrator: throughout his time in Java, he was gathering information to use in the governance of the archipelago and also to make a case for a permanent British colonial presence there. There is a tension between the bureaucratic metanarrative provided by Raffles and his colonial functionaries,

---

**7** See Karen Barad, *Meeting the Universe Halfway: Quantum Physics and the Entanglement of Matter and Meaning* (Durham, NC: Duke University Press, 2007).

**8** A more detailed and wide-ranging account of British Romantic writing and the Tambora eruption can be found in my forthcoming monograph, *British Romanticism, Climate Change, and the Anthropocene: Writing Tambora.*

**9** Charles Assey, "Narrative of the Effects of the Eruption from the Tomboro Mountain, in the Island of Sumbawa, on the 11th and 12th of April 1815," *Transactions of the Batavian Society of Arts and Sciences* 8 (1816): 3-4. Further references are in the text.

and the text's more localized accounts of the catastrophe. Although the report emphasizes the sublime power of the eruption, it also presents itself as a repository of objective environmental knowledge that will support imperial power. In contrast, the individual accounts that it contains are more often shown to invoke confusion and legend. The document, therefore, can be seen as marking a shift from a religious-mythic understanding of disaster in which "morality and materiality, social relations and natural phenomena, were understood to be interrelated" to a more "modern understanding [...] according to which the merely material realm of nature followed its own mechanistic principles that were entirely separate from human morality and social relations."[10] It reveals a nascent and uneven distinction between a "fact-making," supposedly objective and scientific worldview embodied by the colonial metanarrator and the more embedded "meaning-making" individual narratives within it.[11]

It is fitting that the key source of information about the Tambora eruption is a collection of different perspectives. The catastrophe itself, after all, was not a single event. To use Jane Bennett's term, it was an "assemblage": a grouping "of diverse elements, of vibrant materials of all sorts."[12] The word *Tambora* can stand simply for the volcano itself, but it can also be used to refer to this assemblage of energy and matter, which includes the initial explosions, the Plinian eruption, the lava flow, the pyroclastic currents that destroyed villages, the ash fallout, the darkness that covered a large area, the fall of pumice from the sky, the tsunami, and even the subsequent global climate change caused by the release of sulphur into the stratosphere. Bennett argues that the elements that make up an assemblage can include "humans and their (social, legal, linguistic) constructions," as is made evident by her case study of the electrical power grid.[13] An assemblage, too, can have a form of agency due to "the

---

**10** Kate Rigby, *Dancing with Disaster* (Charlottesville: University of Virginia Press, 2015), 3.

**11** I have taken this distinction from Sheila Jasanoff, "A New Climate for Society," *Theory, Culture & Society*, vol. 27, no. 2–3 (May 2010): 233–253.

**12** Jane Bennett, *Vibrant Matter: A Political Ecology of Things* (Durham, NC: Duke University Press, 2010), 23.

**13** Bennett, *Vibrant Matter*, 24.

vitality of the materials that constitute it."[14] Bennett's work and that of other new materialist thinkers thereby problematizes the dichotomy by which human beings are seen either as shapers of nonhuman nature or victims of its whims by focusing instead on "a material-semiotic network of human and nonhuman agents incessantly generating the world's embodiments and events."[15] An uncomfortable question raised by a phenomenon like Tambora is the extent to which humans and animals participate within it rather than simply being acted upon by it. To make such a suggestion in the face of the immense suffering that accompanies an environmental catastrophe on such a scale may seem like a kind of obscenity. And yet a volcanic eruption does not in itself kill anyone. What we understand as an environmental catastrophe or "natural disaster" is a set of interactions between earth systems, organic beings, and their discursive-material constructions.

In documenting these interactions, the Tambora report shows how individual human experience of a large assemblage is inevitably partial and mediated, as evidenced by the silent darkness that descends across a large area after the eruption: an experience that is simultaneously revealing and obscuring. The document's metanarrative attempts to overcome individual partiality by bringing together different perspectives, and correcting them when necessary, but itself inevitably offers a partial view. Most importantly, as a document of empire in the language of the invading country, it tends to obscure indigenous voices, and indigenous people bore the brunt of the volcano's effects. The few indigenous perspectives that we do encounter are mediated by being reported within the accounts of European witnesses and within the document's metanarrative. Whereas we might expect a fictionalized account of such an event to attempt to evoke empathy with the suffering of affected people, perhaps through depictions of interiority, the "Narrative" is mainly concerned with the observations of colonizing subjects, made from a position of relative safety. The apparently empiricist mode of the document is no doubt why it has been so useful for scholars trying to

---

**14** Bennett, *Vibrant Matter*, 34.

**15** Serenella Iovino and Serpil Oppermann, "Introduction," in *Material Ecocriticism*, ed. Iovino and Oppermann (Bloomington and Indianapolis: Indiana University Press, 2014), 1–17, 3.

uncover the history of the eruption, and a basic reading can tell us much about the different physical phenomena that made up the Tambora assemblage. But a more sophisticated analysis needs to recognize that there is also much that it does not tell us, particularly about the human costs of the eruption. The document offers a particular perspective that encodes a kind of cultural politics.[16]

The original account is around 4,500 words in length and titled "Narrative of the Effects of the Eruption from the Tomboro Mountain, in the Island of Sumbawa, On the 11th and 12th of April 1815." It is dated September 28, 1815, and was first published in the eighth volume (1816) of the Anglo-Dutch *Transactions of the Batavian Society of Arts and Sciences.* The narrative is unsigned but is introduced as "*communicated by the [Society's] President.*" Introducing the version that appears in the *History of Java,* Raffles notes that it was "drawn up by Mr. [Charles] Assey" from the responses to Raffles's circular requiring information about the eruption. Assey was at the time the secretary to the British Government in Java. The "Narrative" is sometimes cited as being authored by J. T. Ross, although I am not aware that any scholar has discussed the attribution.[17] The *Java Half-Yearly Almanac and Directory for 1815* lists the Rev. Professor J. T. Ross as minister of the Protestant Church and connected to the government administration in Batavia.[18] The eighth volume of the *Transactions of the Batavian Society* also lists him as the Society's president—he had previously been its secretary—as Raffles had stepped down at some point after learning in September 1815 that the East India Company was dismissing him from his post and that he would be leaving Java. Whoever first cited the "Narrative" as by Ross seems to have assumed that "*communicated by the President*" indicates that he was the author. However, given the date of the document, which is well before Raffles left Java on March 25, 1816, it seems much more likely that the reference is to the

**16** For an excellent discussion that focuses on the cultural politics of the narrative in another publication context, see Gillen D'Arcy Wood, "The Volcano Lover: Climate, Colonialism, and the Slave Trade in Raffles's *History of Java* (1817)," *Journal for Early Modern Cultural Studies*, vol. 8, no. 2 (Fall/Winter 2008): 33–53.

**17** See, for example, Wood, *Tambora*, 239.

**18** The almanac is available online at www.sites.google.com/site/sumatraswestkust/java-almanac-1.

Society's outgoing president rather than its incoming one. And, even if it does refer to Ross, "*communicated*" is likely to mean no more than that it was delivered by him at a Society meeting. There is absolutely no reason to think that Ross had anything to do with its composition or to doubt Raffles's later claim that it was put together by Assey. Previous scholars have not troubled themselves about what may seem a pedantic point. However, the ambiguity about the document's authorship tells us something about its collaborative nature and its bureaucratic stance: a text that is "drawn up" rather than composed or created. Even as it draws on subjective narratives, this impersonality supports the narrative's claim to authoritative and supposedly objective "fact-making."

To understand the document fully, it is useful to know something about the Society. Raffles sought to find out as much as possible about Java once he had become lieutenant-governor in late 1811. He also sought to revive the moribund Batavian Society of Arts and Sciences, which had been founded by the Dutch in 1778.[19] New by-laws were passed in November 1812, and Raffles was elected president in April 1813. Like its Dutch forebear, this new version of the Society had as its principal objective the pursuit of knowledge for the "public benefit" and, although obviously a part of the colonizing process, was not designed to undertake business for the government or the East India Company.[20] Nonetheless, Raffles held the Society's meetings at his government residence and printed its *Transactions* freely using the administration's press, so it had at least a semi-official imprimatur and noted in its new regulations that it was "under the special patronage of the British Government."[21] Raffles felt that a greater understanding of the natural history of the archipelago, and its indigenous languages, would help to develop Java as a productive colony:

---

**19** Some of the information in this paragraph is taken from Eloise Smith Van Niel, *The Batavian Society and Scholarship in the Dutch East Indies, 1778–1850* (MA Thesis, University of Hawaii, May 1980), chapter 3.

**20** "New Regulations for the Batavian Society of Arts and Sciences" are printed in the seventh volume of the *Transactions of the Batavian Society of Arts and Sciences* (1814): xiv–xvii.

**21** "New Regulations," xiv.

> As every untrodden path affords some new incitement to the inquisitive mind, so we may look for much in the various branches of Natural History;—to the philosophic mind a wide and interesting field is opened, and while we contemplate in a political point of view the advantages which must result from bringing forth and directing in a proper course the latent energies and resources of so large a portion of the habitable globe, it must be a pleasing reflection to the Philanthropist that so many of our fellow creatures are thus gradually retrieved from ignorance, barbarism, and self-destruction.[22]

There could hardly be a clearer articulation of the imperial ideal of a beneficial relationship between scientific knowledge, political power, colonial development, and moral progress. It assumes, however, that the area's "latent energies" are capable of being properly directed; one wonders if the Tambora eruption gave Raffles pause for thought about the potential recalcitrance of the Indonesian environment and its peoples to colonial "improvement."

The Tambora narrative therefore fitted in very well with the journal and with the Society's mission of increasing Western knowledge of the archipelago. However, unlike the other articles, it was also an official government document. It comprises an introduction, and then extracts from letters by the residents (East India Company officials) at Grissie, Sumanap, Baniowangie, and Bima, an unnamed person based at Fort Malboro on Sumatra, and from a Lieutenant Owen Phillips, the assistant resident at Macassar, whom Raffles dispatched to Sumbawa on a relief mission. It also contains accounts from "the Noquedah [Captain] of a Malay Prow" (mediated in the third person) passing near Tambora on April 11, the commander of "the Honorable Company's Cruizer Benares" (first person), and the "Honorable Company's Cruizer Teignmouth" (third person). Within these narratives are general comments on indigenous responses as well as a "direct" eyewitness account

22 Thomas Stamford Raffles, "A Discourse Delivered at a Meeting of the Society of Arts and Sciences in Batavia, on the Twenty-fourth day of April 1813, being the Anniversary of the Institution," *Transactions of the Batavian Society of Arts and Sciences* 7 (1814): 26–27. Further citations are indicated in parentheses in the text.

of the eruption by the Rajah of Saugar (the kingdom next to Tambora) that is incorporated within Phillips's narrative. On the one hand, this elaborate structure might suggest the impossibility of offering a single true rendering of any catastrophe. On the other, the implication is that, through its distanced perspective, the imperial center has the capacity to harness these different local accounts and therefore to produce authentic and stable knowledge. This impetus works against Tambora's centrifugal tendency to disperse outward from the site of the eruption. Colonial fact-making is also built on a vast lacuna, which the Rajah's brief and heavily mediated account cannot fill: the meaning-making stories of the indigenous inhabitants of Sumbawa and other islands who experienced the eruption not as a sublime curiosity but as an immediate threat to life and livelihood.

The opening of the narrative is brisk and functional and establishes the document's status as the collaborative product of an imperial bureaucracy rather than of an individual observer:

> To preserve an authentic account of the violent and extraordinary Eruption of the Tomboro Mountain on Sumbawa in April last, the Honorable the Lieutenant Governor required from the several Residents of districts on the Island, a statement of the circumstances that occurred within their knowledge, and from their replies the following Narrative is collected. (3)

The "extraordinary" violence of the eruption is brought into the control of the machinery of colonial government; the word *required* emphasizes that this is knowledge produced by centralized power. The passive voice at the end of the quotation suggests the impersonality of this process. As a result, the narrative is able to offer an overview of the eruption and its effects that supposedly transcends the limitations of the peripheral and localized knowledge offered by the residents: it can therefore claim to be "an authentic account." The entirely ordinary administrative action displayed by the governor's directions, the obedient responses of the residents, and the production and publication of the narrative is produced by the very extraordinariness of the events described. The only way to deal with the catastrophic, the document implies, is to be as measured and orderly as possible.

A notable feature of the document is its listing of different local explanations for the noise of the eruption and for the eruption itself. It is clear that the immediate assumption of anyone hearing the explosions, without experiencing any of the other phenomena, was that guns were being fired. As Raffles's document states, on Java "the noise was in the first instance almost universally attributed to distant cannon; so much so that a Detachment of Troops were marched from Djocjokarta, in the expectation that a neighboring Post was attacked, and along the Coast boats were in two instances dispatched in quest of a supposed ship in distress" (3–4). Cannon-like sounds were also heard in many other places, including Macasser on South Sulawesi: "towards sun-set the reports seem to have approached much nearer and sounded like heavy guns with occasional slight reports in between" (12–13). Assuming pirate activity, "a Detachment of Troops was embarked on board the Honorable Company's Cruizers Benares and sent in search of them" (13). Even when the fall of ashes reveals that a volcano is the source of the noises, the document emphasizes that nobody realizes that the eruption is as far away as Sumbawa: "it was attributed to an Eruption from the Marapi, the Gunung Kloot, or the Gunung Bromo" (4).

These colonial misunderstandings of the source of the noise are reported alongside indigenous readings of the eruption. The Resident at Grissie (in East Java) relates the following response to the darkness and ash fall:

> I am universally told that no one remembers, nor does their tradition record, so tremendous an Eruption—some look upon it as typical of a change, of the re-establishment of the former Government; others account for it in an easy way by reference to the superstitious notions of their legendary tales, and say that the celebrated Nyai Lorok Kidul has been marrying one of her children, on which occasion she has been firing salutes from her supernatural Artillery. They call the ashes the dregs of her Ammunition. (7)

These readings of the eruption as symbolizing political or supernatural events contrast strongly with the scrupulously empirical account by the resident himself (discussed below). The phrase *superstitious notions* suggests his dismissive attitude to such narratives. A little later in the document, the metanarrator reports more neutrally that "the Balinese attributed the event to a recent dispute between the two Rajahs of Bali B'liling, which terminated in the death of the younger Raja by order of his Brother" (9–10). At this point, before the eruption has been discussed in any detail, the indigenous interpretations do not seem any more or less valid than the more rationalistic reports that also misunderstand the eruption's source. The implication of the Raffles account is that the true story of the catastrophe can only emerge from its *post-facto*, totalizing, and supposedly disinterested narrative.

Many cultures have interpreted unusual natural events such as volcanic eruptions as presaging or reflecting a likely political or social change. In the case of Tambora, the inhabitants of east Java were, in fact, correct that the "former Government" would shortly be re-established. Given that the narrative is dated September 28, 1815, we might reasonably assume that it was composed during that month. During 1814 and 1815, the Raffles administration existed in an unstable geopolitical context. Java had been returned to the Dutch by the Anglo-Dutch Treaty of August 1814. A year later, news arrived on Java of Napoleon's escape and resurgence, which potentially voided the treaty and presented the possibility of a longer British presence in Java. And yet in September, Raffles received the devastating news that he was likely to be dismissed from office. That this sense of political instability may have informed the Tambora narrative is apparent in its description of the weather on Java following the initial explosions:

> From the 6th, the sun became observed: it had everywhere the appearance of being enveloped in fog, the weather was sultry and the atmosphere close and still; the sun seemed shorn of its rays, and the general stillness and pressure of the atmosphere foreboded an Earthquake. (4)

This description alludes, rather gracefully, to the famous passage in Book One of *Paradise Lost* describing Satan as a "ruin'd" archangel after his fall:

> As when the Sun new-ris'n
> Looks through the Horizontal misty Air
> Shorn of his Beams, or from behind the Moon

In dim Eclipse disastrous twilight sheds
On half the Nations, and with fear of change
Perplexes Monarchs.[23]

The allusion connects the mist-covered sun to a catastrophic overturning of the normal state of things—as indeed it was—and, more specifically, it connects meteorological phenomena to political ones. While emphasizing the status of the document as a production of elite Western culture, the allusion connects the imperialistic metanarrative to the more localized indigenous accounts that read sudden environmental change as signaling some change in the state of human affairs. It also draws attention to the complex mediations that inform the "Narrative." Individual experiences of the eruption are shown to be mediated through distance, fog, and darkness; eyewitness accounts are mediated through the metanarrator; the metanarrator's perspective is here mediated through the Miltonic sublime; and the narrative is itself mediated through Assey, Raffles, and the different contexts of its publication. The narrative's claims to exert discursive authority from a distanced perspective are to some extent contradicted by its heteroglossic mode and a complex entanglement of stories that are all shown to be part of the Tambora assemblage rather than external to it.

The "Narrative" shows that the Tambora catastrophe occurs not just in the immediate area of the eruption but across the archipelago. An analysis of the eyewitness accounts from different locations reveals its "distributive agency," as well as the ordering process by which its discursive and material aspects come together.[24] But Tambora is not simply multilocational; it is also nonlocational in the sense that it challenges the distinction between different places.[25] After the main eruption in the evening of April 10, a remarkable and terrifying darkness covered a very

**23** John Milton, *Paradise Lost*, ed. Christopher Ricks (London: Penguin, 1968), 22 (I.593–599).

**24** Bennett, *Vibrant Matter*, 21.

**25** I have taken the term *nonlocality* from Timothy Morton but use it in a slightly different way from him, in part because he is keen to distinguish hyperobjects from assemblages: see *Hyperobjects: Philosophy and Ecology after the End of the World* (Minneapolis: University of Minnesota Press, 2013): 2.

large area. It is described with particular precision by the resident at Grissie:

> I woke in the morning of the 12th, after what seemed to be a very long night, and taking my watch to the lamp found it to be half past eight o'clock, I immediately went out and found a cloud of ash descending; at 9 o'clock no day light—the layer of ashes on the terrace before my door at the Kradenan measures one line in thickness; ten AM—a faint glimmering of light can now be perceived overhead: half past 10—can distinguish objects 50 yards distant: 11 AM—Breakfasted by Candle-light, the birds began to chirrup as at the approach of day: half past 11—can discover the situation of the sun through a thick cloud of ashes; 1 PM found the layer of ashes one line and a half thick, and measured in several places with the same results; 3 PM the ashes have increased one eighth of a line more; 5 PM it is now lighter, but still I can neither read nor write without Candles. (6–7)

Crucial to this passage is the precise and minute demarcation of time and space: the kind of precision that you would find in a sea captain's log. In the absence of the usual temporal markers of day and night, imperial clock time becomes a vital way of making sense of what is happening and finding order in disorder. A "line" is an old British measurement, usually reckoned at one-twelfth of an inch. Measuring the depth of the layer of ashes can therefore be understood as particularly precise, as the length increases from around 2 millimeters to 2.5 millimeters to 2.75 millimeters. As one would expect of a sensible colonial official, the writer does not panic but proceeds to make orderly observations of the unusual phenomenon. The colonizing subject, therefore, is shown to have the knowledge and technology to comprehend even the strangest and most confusing local conditions, to bring a little light into the darkness. And yet this comprehension becomes meaningful only within an assemblage of other partial accounts mediated by a disinterested central intellect.

Perhaps in order to maintain the illusion of disinterestedness, it is only in the document's last four pages, in Lieutenant Phillips's account from Sumbawa, that readers are given some sense of the eruption's more direct effects on the local population.

Phillips was apparently "well versed in the Malayan language," and his colonial account is the only one to include anything like a proper engagement with an indigenous perspective on the eruption.[26] Each paragraph of his account is in quotation marks, suggesting that his words are reproduced verbatim. Sent by Raffles to manage the delivery of "a supply of rice to their relief" and to discover the "local effects," Phillips notes that

> The extreme misery to which the inhabitants have been reduced is shocking to behold—there were still on the road side the remains of several corpses and the marks of where many others had been interred—the Villages almost entirely deserted—and the houses fallen down—the surviving inhabitants having dispersed in search of food. (21–22)

In a narrative so lacking in emotion, *extreme* and *shocking* have a powerful charge. This sense of total societal collapse continues as Phillips notes that "a violent Diarrhoea [...] has carried off a great number of people" (22). The severity of the famine is further emphasized by the report of the Rajah of Saugar that "one of his own daughters died from hunger." The Rajah's voice is in the third person and is so far removed from the narrative itself, mediated through Phillips (and through Assey and Raffles), that there is little sense of him as a character to be sympathized with. Phillips notes that the Rajah "was himself a spectator of the late Eruption" and therefore that his account "is perhaps more to be depended upon than any other I can possibly obtain" (23). The three paragraphs that follow give the only known eyewitness account of the initial eruption itself, and for two of them, which focus on description, we might assume that we are getting a composite of Phillips's and the Rajah's voices. The return to Phillips's colonial perspective is apparent toward the end of the third paragraph, which notes that "the Natives are apprehensive of another Eruption during the ensuing rainy season" (24).

The Rajah's account is the most dramatic section of the "Narrative" and gives a sense of the extreme power of an

---

**26** John Crawfurd, *History of the Indian Archipelago*, 3 vols. (Edinburgh: Constable, 1820), II, 124.

elemental assemblage beyond human control:

> The fire and columns of flame continued to rage with unabated fury [...]. Stones at this time fell very thick at Saugar—some of them as large as two fists, but generally not larger than walnuts; between 9 and 10 PM ashes began to fall and soon after a violent whirlwind ensued, which blew down nearly every house in the village of Saugar, carrying the tops and light parts away with it. In the part of Saugur adjoining Tomboro, its effects were much more violent, tearing up by the roots the highest trees, and carrying them into the air together with men, houses, cattle and whatever else came within its influence (this will account for the large number of floating trees seen at sea). The sea rose nearly 12 feet higher than it had ever been known to be before, and completely spoiled the only small spots of rice lands in Saugur—sweeping away houses and everything within its reach. (23–24)

The passage underscores human vulnerability to elemental forces, moving through fire, earth, air, and water. The idea of the fire "raging" and in "fury" is more anthropomorphic than the "Narrative" usually gets and suggests a certain agency behind this conflagration. It might be linked to the more celebratory personification earlier in the document, where it is reported that the eruption was caused by the deity "Nyai Loroh Kidul" marking the marriage of one of her children. As in other parts of the narrative, the description of abnormal phenomena is an attempt to exert representational control through the demarcation of time and other measurements (the size of the stones and the height of the sea), as well as grammatical and syntactical ordering, but here it only seems futile in the context of elemental violence and mortal danger. The destruction of human dwellings and means of sustenance, often so important to representations of catastrophe, is emphasized when the narrative shifts back to Phillips's overview, as he notes that only one of the villages in Tomboro is remaining and that "in Precate, no vestige of a house is left" (24). He estimates there were "certainly not fewer than 12,000 individuals in Tomboro and Precate at the time of the Eruption" (24–25), the implication being that the majority of them have perished. However, this suggestion is not developed and is in any case a vast underestimate of the catastrophe's local death toll. Phillips goes on to state that a "high point" near the eruption avoided the complete destruction of "trees and herbage" along the "North and West sides of the Peninsula" and that three people were saved on the night of the eruption, although he has not been able to find them (25). This "search" for the three survivors inevitably exists in an ironic contrast with the 12,000 people mentioned in the previous paragraph and who Phillips believes have been destroyed.

The narrative attempts to end on a hopeful note. In response to the report from a messenger to Sumbawa that "an immense number of people have been starved," Phillips notes, in the document's final sentence, that "the distress has, however, I trust, been alleviated by this time, as the brig with 63 coyangs of Rice from Java arrived there the day he was leaving it" (25). As Wood points out, this amount of aid would have lasted only a few days—although to be fair, Raffles and Phillips probably had little idea of the scale of the devastation.[27] (The awkward parenthetical caveats "however, I trust" may suggest Phillips's lack of certainty on this score.) Earlier in the document, the metanarrator notes that the lieutenant governor had "dispatched a supply of rice" for the inhabitants' relief, under Phillips's command (11). The "Narrative" ends, therefore, with a reminder of the supervening intelligence and care of the British administration, which observes unusual phenomena disinterestedly and dispassionately but also exercises its duty to the indigenous peoples within its sphere of influence. The first publication of the "Narrative" within the journal of a society dedicated to producing knowledge that would support colonial development is therefore entirely fitting and suggests that it does much more than simply provide a neutral account of the eruption, despite the uses to which it has generally been put by scholars. Its heteroglossic form, rather, marks it out as part of the discursive material Tambora assemblage: an imbrication of living beings, cultural constructions, energy flows, and objects. That is, the narrative is not merely a commentary on a political-ecological catastrophe but an intervention within it.

---

**27** Wood, *Tambora*, 31.

## Byron, the Shelleys, and the End of the World

During the summer of the 1816, Lord Byron, Percy Bysshe Shelley, and Mary Shelley were sojourning in Switzerland, where they wrote and conceived some of the most important texts in the British Romantic canon. Scholars have noted the importance of the global climate crisis to works such as Byron's "Darkness" (1816), Percy Shelley's "Mont Blanc" (1817), and Mary Shelley's *Frankenstein* (1818). However, this insight has tended to generate broad contextual readings rather than close analysis. This section, focusing on *Frankenstein*, sheds new light on the complexity with which Byron and the Shelleys wrote about environmental catastrophe in 1816. While they knew nothing of Tambora, the bad weather that it largely caused was an important influence on their creativity, in combination with their interest in contemporary natural philosophy and their experience of the sublime landscapes around Geneva. The wider project from which my discussion drives brings together their 1816 writings to reveal the richness of their reflections on the vulnerability of human communities living with uncontrollable geophysical and climatic forces, the entanglement of humans and nonhuman nature, and the possibility of human extinction. A key influence on their thought was the French natural philosopher the Comte de Buffon, whose "Les époques de la nature" (1778) suggested that the Earth had undergone a process of gradual cooling since its creation and imagined an icy future in which it would be rendered uninhabitable.[28]

One of the key moments in *Frankenstein* is set around Mont Blanc, which had a powerful impact on all three authors' works. In the novel, after the trauma of William's murder by the "creature" and Justine's unjust execution, the Frankenstein family hope to recuperate on "an excursion to the valley of Chamounix."[29] One morning, Victor awakens to depressingly bad weather, reminiscent of the "dreary night of November" when he brought the Creature to life in what is described as a "catastrophe" (84).

---

**28** Martin Rudwick, *Bursting the Limits of Time: The Reconstruction of Geohistory in the Age of Revolution* (Chicago: University of Chicago Press, 2005), 142–149.

**29** Mary Shelley, *Frankenstein*, ed. D. L. Macdonald and Kathleen Scherf, 2nd ed. (Peterborough, Ontario: Broadview, 2005), 121. Further citations are indicated in parentheses in the text.

He seeks solace in a lone trek to the summit of Montanvert, on the northern slopes of Mont Blanc. "The presence of another would destroy the solitary grandeur" of the landscape, he notes,

> It is a scene terrifically desolate. In a thousand spots the traces of the winter avelanche may be perceived, where trees lie broken and strewed on the ground [...] The path, as you ascend higher, is intersected by ravines of snow, down which stones continually roll from above; one of them is particularly dangerous, as the slightest sound, such as even speaking in a loud voice, produces a concussion of air sufficient to draw destruction upon the head of the speaker. (123)

Like other alpine tourists of the period, Victor seeks an encounter with the sublime. But this version of the experience does not involve the safe contemplative distance that is often associated with the ability to aestheticize a potentially dangerous landscape. As befits Victor's risk-taking character, he emphasizes the physical danger affecting the human observer. The "slightest sound" in this setting needs to be avoided: the silence necessary to avoid causing an avalanche also accentuates the alienness of the landscape. This emphasis on silence connects the novel to Percy Shelley's poem "Mont Blanc," as does the threat to human dwelling represented metonymically through the destruction of the trees. Along with glacial augmentation, this is a key trope in the repertoire of Byron and the Shelleys in 1816. For example, in Mary's journal on July 24, 1816—when she also makes the first mention of composing *Frankenstein*—she writes that:

> Nothing can be more desolate than the ascent of this mountain—the trees in many places have been torn away by avelanches and some half leaning over others intermingled with stones present the appearance of a vast & dreadful desolation.[30]

---

**30** *The Journals of Mary Shelley, 1814–44*, ed. Paula R. Feldman and Diana Scott-Kilvert, 2 vols. (Oxford: Clarendon Press, 1987), I, 117–118.

The Shelleys' prose accounts of the pines stress the scene's superlative "desolation"; that is, its barrenness and its lack of inhabitants. *Desolate* derives from the Latin *desolatus*, meaning "left alone": this is a landscape in which human communities cannot flourish.

After ascending Montanvert, Victor walks across the glacier to the opposite mountain so that he has a view of Mont Blanc "in awful majesty" (124). The individual apotheosis potentially offered by the sublime—the swelling of the heart provoked by a "wonderful and stupendous scene"—is cut short by his sudden encounter with the Creature, their first since his creation (124). Victor's sublime solitude is interrupted by a painful reminder of the communal responsibilities one has to others but also of his inferiority to his creation, who travels with "super-human speed" and "easily eludes" Victor's attempt to attack him physically. The Creature responds to his creator's loathing with counterthreats, but also with justifications, pointing out his natural benevolence, his loneliness, and Victor's failure to care for him. Therefore, he states,

> The desert mountains and dreary glaciers are my refuge. I have wandered here many days; the caves of ice, which I only do not fear, are a dwelling to me, and the only one which man does not grudge. These bleak skies I hail, for they are kinder to me than your fellow-beings. (126–127)

The inhospitable landscape where humans and animals cannot "dwell," and which for Victor is a kind of touristic site, is for the Creature the safest dwelling that he can find. Cast out by his creator, and by the human communities to which his sensibility, if not his appearance, should connect him, the Creature has been forced to find a different sort of connection by living within the apparently "dead" world of rocks and glaciers and finding a form of fellowship (as the punning word "hail" suggests) even with the "bleak" weather.

The anthropologist Julie Cruikshank has contrasted a Western colonial idea of glaciers as "pristine, wild, and remote from human influence" with the views of indigenous populations in Alaska and the Yukon, who (she suggests) see them as "intensely social spaces where human behavior, especially casual hubris or arrogance, can trigger dramatic and unpleasant consequences in the physical world."[31] Whereas for Victor the glacier is a sublime backdrop to his anthropocentric imagination, for the Creature it offers the possibility of a relationship with the world. Despite his apparently "unnatural" beginnings, the Creature is therefore shown here and at other points in the text to be more connected to nonhuman nature than to the human characters. He can flourish in any environment but has to tell his story to Victor in a constructed dwelling: a mountain hut with a fire and therefore at a temperature suitable for his creator's "fine sensations" (127). Nonhuman nature is presented in idyllic, Rousseauvian terms, as when he promises that he "will go to the vast wilds of South America" with his mate and live a vegetarian existence: "we shall make our bed of dried leaves; the sun will shine on us as on man and will ripen our food. The picture I present to you is peaceful and human" (170). But writers of the Romantic period were increasingly aware of another state of nature: a state of rapid, violent, and uncontrollable change. The encounter between Victor and the Creature in the vale of Chamonix connects the Creature's agency to that of the glaciers. This agency is often destructive, although the allusion to Samuel Taylor Coleridge's "Kubla Khan" (the "caves of ice" within the Khan's "pleasure-dome") in the Creature's speech may suggest a creative power as well. After all, in Percy Shelley's "Mont Blanc," the glaciers not only destroy dwellings but also construct "dome, pyramid, and pinnacle." Furthermore, beneath this "flood of ruin,"

> [...] vast caves
> Shine in the rushing torrent's restless gleam,
> Which from those secret chasms in tumult welling
> Meet in the vale, and one majestic River,
> The breath and blood of distant lands, for ever
> Rolls its loud waters to the ocean waves,
> Breathes its swift vapours to the circling air.[32]

---

31 Julie Cruikshank, *Do Glaciers Listen? Local Knowledge, Colonial Encounters, and Social Imagination* (Vancouver: University of British Columbia Press, 2005), 10–11.

32 Percy Bysshe Shelley, "Mont Blanc," in Mary Shelley and Percy Bysshe Shelley, *History of a Six Weeks Tour* (London: T. Hookham and C. and J. Ollier, 1817), 181–182.

The allusions to "Kubla Khan" are palpable, but whereas in Coleridge's poem, the river that bursts forth from "that deep romantic chasm" eventually sinks "in tumult to a lifeless ocean," here the glaciers are seen as part of a global hydrological cycle upon which life is dependent. It is the Creature's potential to create new life that Victor finds most disturbing.

The Creature's power and resilience generates reflections on the future of humanity as a species. His appearance offers an ironic commentary on Victor's fantasy on the way to Chamonix of "the mighty Alps, whose white and shining pyramids and domes towered above all, as belonging to another earth, the habitations of another race of beings" (121). In a recent discussion of *Frankenstein* and human extinction, Claire Colebrook follows earlier political readings of the Creature as "a disenfranchised other who could, in theory, be redeemed and included."[33] She places Mary Shelley with thinkers like Marx, Adorno, and Jameson, who seek political solutions to what may seem as an intolerable existence: "what appears to be existentially unacceptable should be transformed through social and political revolution. If recognition were granted to the potential hordes of the future one would be faced not with violence but with sympathy and pity."[34] Colebrook seems to see such thinking as narrow and utopian, given humanity's impact on the environment: "the question is not one of how we humans can justify hostile life, but how we can possibly justify ourselves given our malevolent relation to life."[35] As Colebrook recognizes, political critique is crucial to the novel; nonetheless, she underestimates its willingness to face up to the prospect of human extinction rather than seeing it as a problem to be solved. The Creature may present to Victor a vision of a "peaceful and human" existence in South America with his mate (170), but this is a rhetorical ploy; to view the Creature as no more than a surrogate for disenfranchised humans is to miss precisely what is important about the novel in the context of Romantic geology and the Villa Diodati group's concern with human-environmental

---

**33** Claire Colebrook, *Death of the PostHuman: Essays on Extinction, Vol. 1* (Ann Arbor, MI: Open Humanities Press, 2014), 195.

**34** Colebrook, *Death of the PostHuman*, 196.

**35** Colebrook, *Death of the PostHuman*, 198.

interactions. Victor eventually destroys his work on the Creature's companion due to his fear that the two might procreate: "a race of devils would be propagated upon the earth, who might make the very existence of the species of man a condition precarious and full of terror" (190). Victor assumes that this new posthuman species would be far more resilient than humans and would therefore destroy them; having read Buffon, he is well aware of the adaptive value of being able to survive in colder climates. It is therefore entirely fitting that the novel begins and ends in the Arctic. After being responsible for the deaths of Victor's family, the Creature leads him "to the everlasting ices of the north" so that he will suffer further privations: "you will feel the misery of cold and frost, to which I am impassive" (227). Frankenstein raises the specter of a posthuman future in which a new species develops that is able to flourish on Buffon's icy globe.

As Siobhan Carroll has recently shown, Arctic exploration in the Romantic period was enmeshed with debates around climate change and geoengineering projects. She argues convincingly that "situating Frankenstein in a climatological context enables us to see the ramifications of Victor's experiment as symptomatic of a larger cultural concern over Europeans' readiness to wield the nature-shaping power of imperial science."[36] The relationship between knowledge and power is a key concern of the novel and evident in how Walton and Victor try to impose their wills onto the world. One of Walton's goals is to find the fabled Northwest Passage linking the Atlantic to the Pacific. Such a discovery would have been of enormous benefit to British imperialism and was the aim of a number of government-supported expeditions in the nineteenth century. Walton's "voyage of discovery" through "pathless seas" (58, 317) is an attempt at reterritorialization: by beginning with an entirely deterritorialized idea of the Arctic as a blank space without indigenous human inhabitants, nonhuman creatures, or elemental agencies, he presents the possibility that it can be mastered cartographically as well as physically so that he may achieve "the sight of a part of the world never before

---

**36** Siobhan Carroll, "Crusades Against Frost: *Frankenstein*, Polar Ice, and Climate Change in 1818," *European Romantic Review*, vol. 24, no. 2 (March 2013): 211–230, 220.

visited, and may tread a land never before imprinted by the foot of man" (50). His fantasy is of a land "where cold and frost are banished" (49), but the Arctic itself proves resistant to imaginative projections: "Frankenstein sounds the death knell of dreams of the undiscovered *terra nullius* in its depiction of Walton's defeat in polar space." This resistance is most powerfully apparent in the cold reality that "immures" his ship among threatening "mountains of ice" (234) and forces Walton and his crew to turn back when the ice eventually breaks up. Imperialism seeks to impose itself onto the environment through writing the landscape as possession. However, in Shelley's novel, ice is a force recalcitrant to any such process. Furthermore, their experience of the alpine landscape and Buffon's speculations suggested to the Shelleys that the empire of ice itself had the capacity to overwrite all traces of the human.

The novel's concern with Arctic exploration is inflected by its relationship to "The Rime of the Ancyent Marinere." In his second letter to his sister, Walton notes that "I am going to unexplored regions, to 'the land of mist and snow'; but I shall kill no albatross, therefore do not be alarmed for my safety" (55).[37] The poem had been reprinted in 1817, and its depiction of human vulnerability in the face of extreme weather conditions must have resonated particularly during a period of climate crisis: "And now the storm-blast came, and he / Was tyrannous and strong."[38] Like Frankenstein, Coleridge's poem is particularly concerned with the power of ice over humans:

> Listen, Stranger! Mist and Snow,
> And it grew wond'rous cold:
> And Ice mast-high came floating by,
> As green as emerald.
>
> And thro' the drifts the snowy clifts
> Did send a dismal sheen;

---

**37** See Samuel Taylor Coleridge, *Coleridge's Poetry and Prose*, ed. Nicholas Halmi, Paul Magnuson, and Raymond Modiano (New York: Norton, 2004), 68.

**38** Coleridge, *Poetry and Prose*, 63.

Ne shapes of men ne beasts we ken—
The Ice was all between.

The Ice was here, the Ice was there,
The Ice was all around:
It crack'd and growl'd, and roar'd and howl'd—
Like noises of a swound![39]

This is another "desolate" space, where there is no place for humans or animals: another space defined by absence ("ne shapes [...] we ken"). In such an environment, the ice itself is given an uncanny agency through anaphora and personification. Like Mont Blanc's glaciers, it is not static but threatens through its unpredictable movements. The final simile seems to suggest that it resembles a person experiencing some sort of violent fit ("swound"). Coleridge's poem, like *Frankenstein*, presents the failure and vulnerability of humanity's imperializing aspirations when faced with uncooperative objects. Both texts have been subject to dubious interpretations by ecological writers. Bruno Latour finds in Shelley's novel the injunction "love your monsters": "Dr. Frankenstein's crime was not that he invented a creature through some combination of hubris and high technology, but rather that he abandoned the creature to itself." He sees the novel as offering a "parable for political ecology" by showing that human beings need to take responsibility for the entanglements of their technologies with the natural world.[40] Frankenstein is therefore co-opted in line with ecomodernist discourse, which calls for a "good Anthropocene" created by a doubling down on the human capacity to shape the planet.[41] A more traditional ecological moral is evident in James McKusick's account of Coleridge's poem: "by blessing the water-snakes, the Mariner is released from his state of alienation from nature [...] [He] has learned what the Albatross came to teach him: that he must cross the

---

**39** Coleridge, *Poetry and Prose*, 62.

**40** Bruno Latour, "Love Your Monsters: Why We Must Care for Our Technologies As We Do Our Children," *Breakthrough Journal* 2 (Winter 2012), www.thebreakthrough.org/index.php/journal/past-issues/issue-2/love-your-monsters.

**41** See "An Ecomodernist Manifesto," http://www.ecomodernism.org. Latour is not a signatory of the manifesto, although it clearly draws on some of his ideas.

boundaries that divide the natural world, through unmotivated acts of compassion between 'man and bird and beast.'"[42] One might well sympathize with these readings, but they share two problems. First, they focus on the relationship between humans and nonhuman creatures and ignore the significance of recalcitrant objects to the moral framework of both texts. Secondly, they do not reflect the complexity of their sources, and particularly the way in which both texts are concerned with fear, abjection, disgust, and the deep difficulties of connecting with others. Rather than offering straightforward formulas for dwelling in the Anthropocene, *Frankenstein* and the "Rime" are focused on the potential incompatibility of human aspirations and desires with the dynamic environmental processes upon which we are dependent. In different ways, Latour and McKusick offer an inadequately anthropocentric response to texts that, whatever else they do, tend to emphasize the precarity and finitude of the human.

However, addressing the challenge of environmental catastrophe for the human species does not require viewing humanity as an undifferentiated monolith. There may be an instructive parallel between the aestheticized view of the climate crisis taken by sheltered European elites in 1815–18 and the complacent attitude toward present-day global warming taken by many in the West. For Raffles, the Tambora eruption was extraordinary but could be controlled and mediated through the apparatus of colonialism. Similarly, the complex accounts of human precarity in the writings of Byron and the Shelleys were contingent on the authors themselves being relatively sheltered from the climate crisis. Then as now, the people most directly affected by environmental change were not well served by canonical and institutionalized forms of representation and mediation. If a new materialist approach is going to help us to understand the Anthropocene, we need to find ways of analyzing assemblages that shed light on the political, social, cultural, and economic inequalities associated with climate change rather than flattening them through a model of agency that does not attend sufficiently to how power is distributed. This essay has attempted

42 James McKusick, *Green Writing: Romanticism and Ecology* (Basingstoke, UK: Palgrave, 2000), 47.

to make the case for a historicist criticism that is sensitive to the interactions of matter, discourse, and power, and yet no piece of criticism can do justice to the devastating effects of the Tambora eruption and the suffering and death that it caused. To some extent, therefore, my analysis cannot help but rehearse the apparently disinterested, imperializing mode of Raffles's narrative. But I hope that it has also resisted that centralizing urge by addressing Tambora as a material-textual assemblage dispersed not only spatially but also temporally. Taking a genealogical approach to the Anthropocene allows us to produce richer and more inclusive narratives of past political-environmental catastrophes, and these new narratives may themselves help us to reframe the present.

## LETTERS FROM ROBERT WALTON TO MARGARET SAVILLE

Excerpted from *Frankenstein: or, The Modern Prometheus* (Boston: Cornhill Publishing Co., 1922).

Mary Shelley

---

### LETTER I

To Mrs. Saville, England

St. Petersburgh, Dec. 11th, 17—

You will rejoice to hear that no disaster has accompanied the commencement of an enterprise which you have regarded with such evil forebodings. I arrived here yesterday, and my first task is to assure my dear sister of my welfare and increasing confidence in the success of my undertaking.

I am already far north of London, and as I walk in the streets of Petersburgh, I feel a cold northern breeze play upon my cheeks, which braces my nerves and fills me with delight. Do you understand this feeling? This breeze, which has travelled from the regions towards which I am advancing, gives me a foretaste of those icy climes. Inspirited by this wind of promise, my daydreams become more fervent and vivid. I try in vain to be persuaded that the pole is the seat of frost and desolation; it ever presents itself to my imagination as the region of beauty and delight...

I feel my heart glow with an enthusiasm which elevates me to heaven, for nothing contributes so much to tranquillize the mind as a steady purpose—a point on which the soul may fix its intellectual eye. This expedition has been the favourite dream of my early years. I have read with ardour the accounts of the various voyages which have been made in the prospect of arriving at the North Pacific Ocean through the seas which surround the pole...

Farewell, my dear, excellent Margaret. Heaven shower down blessings on you, and save me, that I may again and again testify my gratitude for all your love and kindness.

Your affectionate brother,
R. Walton

* * *

* * * *

LETTER II
To Mrs. Saville, England

Archangel, 28th March, 17—

How slowly the time passes here, encompassed as I am by frost and snow! Yet a second step is taken towards my enterprise. I have hired a vessel and am occupied in collecting my sailors; those whom I have already engaged appear to be men on whom I can depend and are certainly possessed of dauntless courage.

But I have one want which I have never yet been able to satisfy, and the absence of the object of which I now feel as a most severe evil. I have no friend, Margaret: when I am glowing with the enthusiasm of success, there will be none to participate my joy; if I am assailed by disappointment, no one will endeavour to sustain me in dejection. I shall commit my thoughts to paper, it is true; but that is a poor medium for the communication of feeling. I desire the company of a man who could sympathize with me, whose eyes would reply to mine. You may deem me romantic, my dear sister, but I bitterly feel the want of a friend. I have no one near me, gentle yet courageous, possessed of a cultivated as well as of a capacious mind, whose tastes are like my own, to approve or amend my plans. How would such a friend repair the faults of your poor brother!...It is true that I have thought more and that my daydreams are more extended and magnificent, but they want (as the painters call it) *keeping*; and I greatly need a friend who would have sense enough not to despise me as romantic, and affection enough for me to endeavour to regulate my mind...

Well, these are useless complaints; I shall certainly find no friend on the wide ocean...

I cannot describe to you my sensations on the near prospect of my undertaking. It is impossible to communicate to you a conception of the trembling sensation, half pleasurable and half fearful, with which I am preparing to depart. I am going to unexplored regions, to 'the land of mist and snow,' but I shall kill no albatross; therefore do not be alarmed for my safety or if I should come back to you as worn and woeful as the 'Ancient Mariner.' You will smile at my allusion, but I will disclose a secret. I have often attributed my attachment to, my passionate enthusiasm for, the dangerous mysteries of ocean to that production of the most imaginative of modern poets. There is something at work in my soul which I do not understand. I am practically industrious—painstaking, a workman to execute with perseverance and labour—but besides this there is a love for the marvellous, a belief in the marvellous, intertwined in all my projects, which hurries me out of the common pathways of men, even to the wild sea and unvisited regions I am about to explore...

Remember me with affection, should you never hear from me again.

Your affectionate brother,
Robert Walton

* * *

* * * *

LETTER III

To Mrs. Saville, England

July 7th, 17—

My dear Sister,

I write a few lines in haste to say that I am safe—and well advanced on my voyage. This letter will reach England by a merchantman now on its homeward voyage from Archangel; more fortunate than I, who may not see my native land, perhaps, for many years. I am, however, in good spirits: my men are bold and apparently firm of purpose, nor do the floating sheets of ice that continually pass us, indicating the dangers of the region towards which we are advancing, appear to dismay them. We have already reached a very high latitude; but it is the height of summer, and although not so warm as in England, the southern gales, which blow us speedily towards those shores which I so ardently desire to attain, breathe a degree of renovating warmth which I had not expected.

No incidents have hitherto befallen us that would make a figure in a letter. One or two stiff gales and the springing of a leak are accidents which experienced navigators scarcely remember to record, and I shall be well content if nothing worse happen to us during our voyage...

Heaven bless my beloved sister!

R.W.

* * *

* * * *

## LETTER IV
To Mrs. Saville, England

August 5th, 17—

So strange an accident has happened to us that I cannot forbear recording it, although it is very probable that you will see me before these papers can come into your possession.

Last Monday (July 31st) we were nearly surrounded by ice, which closed in the ship on all sides, scarcely leaving her the sea-room in which she floated. Our situation was somewhat dangerous, especially as we were compassed round by a very thick fog. We accordingly lay to, hoping that some change would take place in the atmosphere and weather.

About two o'clock the mist cleared away, and we beheld, stretched out in every direction, vast and irregular plains of ice, which seemed to have no end. Some of my comrades groaned, and my own mind began to grow watchful with anxious thoughts, when a strange sight suddenly attracted our attention and diverted our solicitude from our own situation. We perceived a low carriage, fixed on a sledge and drawn by dogs, pass on towards the north, at the distance of half a mile; a being which had the shape of a man, but apparently of gigantic stature, sat in the sledge and guided the dogs. We watched the rapid progress of the traveller with our telescopes until he was lost among the distant inequalities of the ice.

This appearance excited our unqualified wonder. We were,

as we believed, many hundred miles from any land; but this apparition seemed to denote that it was not, in reality, so distant as we had supposed. Shut in, however, by ice, it was impossible to follow his track, which we had observed with the greatest attention...

In the morning, however, as soon as it was light, I went upon deck and found all the sailors busy on one side of the vessel, apparently talking to someone in the sea. It was, in fact, a sledge, like that we had seen before, which had drifted towards us in the night on a large fragment of ice. Only one dog remained alive; but there was a human being within it whom the sailors were persuading to enter the vessel. He was not, as the other traveller seemed to be, a savage inhabitant of some undiscovered island, but a European. When I appeared on deck the master said, 'Here is our captain, and he will not allow you to perish on the open sea.'

On perceiving me, the stranger addressed me in English, although with a foreign accent. 'Before I come on board your vessel,' said he, 'will you have the kindness to inform me whither you are bound?'

You may conceive my astonishment on hearing such a question addressed to me from a man on the brink of destruction and to whom I should have supposed that my vessel would have been a resource which he would not have exchanged for the most precious wealth the earth can afford. I replied, however, that we were on a voyage of discovery towards the northern pole.

Upon hearing this he appeared satisfied and consented to come on board...

The stranger has gradually improved in health but is very silent and appears uneasy when anyone except myself enters his cabin. Yet his manners are so conciliating and gentle that the sailors are all interested in him, although they have had very little communication with him. For my own part, I begin to love him as a brother, and his constant and deep grief fills me with sympathy and compassion. He must have been a noble creature in his better days, being even now in wreck so attractive and amiable. I said in one of my letters, my dear Margaret, that I should find no friend on the wide ocean; yet I have found a man who, before his spirit had been broken by misery, I should have been happy to have possessed as the brother of my heart.

August 13th, 17—

My affection for my guest increases every day...He is now much recovered from his illness and is continually on the deck, apparently watching for the sledge that preceded his own. Yet, although unhappy, he is not so utterly occupied by his own misery but that he interests himself deeply in the projects of others...I was easily led by the sympathy which he evinced to use the language of my heart, to give utterance to the burning ardour of my soul, and to say, with all the fervour that warmed me, how gladly I would sacrifice my fortune, my existence, my every hope, to the furtherance of my enterprise. One man's life or death were but a small price to pay for the acquirement of the knowledge which I sought, for the dominion I should acquire and transmit over the elemental foes of our race. As I spoke, a dark gloom spread over my listener's countenance. At first I perceived that he tried to suppress his emotion; he placed his hands before his eyes, and my voice quivered and failed me as I beheld tears trickle fast from between his fingers; a groan burst from his heaving breast. I paused; at length he spoke, in broken accents: 'Unhappy man! Do you share my madness? Have you drunk also of the intoxicating draught? Hear me; let me reveal my tale, and you will dash the cup from your lips!'

Even broken in spirit as he is, no one can feel more deeply than he does the beauties of nature. The starry sky, the sea, and every sight afforded by these wonderful regions seem still to have the power of elevating his soul from earth. Such a man has a double existence: he may suffer misery and be overwhelmed by disappointments, yet when he has retired into himself, he will be like a celestial spirit that has a halo around him, within whose circle no grief or folly ventures.

Will you smile at the enthusiasm I express concerning this divine wanderer?...

August 19th, 17—

Yesterday the stranger said to me, 'You may easily perceive, Captain Walton, that I have suffered great and unparalleled misfortunes. I had determined at one time that the memory of these evils should die with me, but you have won me to alter my determination. You seek for knowledge and wisdom, as I once did; and I ardently hope that the gratification of your wishes may not be a serpent to sting you, as mine has been. I do not know that the relation of my disasters will be useful to you; yet, when I reflect that you are pursuing the same course, exposing yourself to the same dangers which have rendered me what I am, I imagine that you may deduce an apt moral from my tale, one that may direct you if you succeed in your undertaking and console you in case of failure. Prepare to hear of occurrences which are usually deemed marvellous. Were we among the tamer scenes of nature I might fear to encounter your unbelief, perhaps your ridicule; but many things will appear possible in these wild and mysterious regions which would provoke the laughter of those unacquainted with the ever-varied powers of nature; nor can I doubt but that my tale conveys in its series internal evidence of the truth of the events of which it is composed.'...

He then told me that he would commence his narrative the next day when I should be at leisure...Strange and harrowing must be his story, frightful the storm which embraced the gallant vessel on its course and wrecked it—thus!

# Cloud Walking

Dehlia Hannah

Charles Stankievech, *LOVELAND*, 2009–2011. HD video still, from installation at Musée d'art contemporain de Montréal, 2011. Courtesy of the artist.

> "Our situation was somewhat dangerous, especially as we were compassed round by a very thick fog."
>
> —Roger Walton, August 5, 17–[1]

Across a great icy expanse, a cloud gathers on the horizon and rushes forward on a gust of wind, filling the air with a dense lavender mist. Engulfed in its shifting currents, I twist my body as though caught in a gale, straining to find a resting place for my gaze. Instinctively restricting my breath as I try to discern the cloud's source and contents, I bend toward the temptation to lose myself in it completely. On the brink of suffocation, it rushes past, vanishing over my shoulder. I turn to follow it, yearning for the return of this fleeting mirage. By turns enthralling and bewildering, the purple cloud holds open a space between the whiteness of the sky and frozen ground. An ambient electronic soundscape gives the tingling impression of ice particles blowing across the ground before lightening in tone to an ethereal din, dropping to a rumble as the cloud circles over the horizon again in an infinite loop. Dazed, apprehensive, entranced, I brace myself to endure the repetition, letting the color seep into my eyes and exhaust my retina until the scene is suffused with the glow of a yellow afterimage. Waves of intoxication, endangerment, delirium conjure associations of smoke and toxic gases.

---

1 Mary Shelley, *Frankenstein: or, The Modern Prometheus*, 1831 edition (Toronto: Broadview Press, 1999), 56.

Each time the rumbling begins anew, I anticipate the crescendo more eagerly. Squinting, I search for a burgeoning shadow in the distance, now tracking its reappearance over the horizon like an astronomical body. The rising sound becomes ominous. Turning in the wind, I hasten to leave suddenly, rushing out of the room before the cloud surrounds me again, before I'm seduced once more into the portentous mass gathering on the horizon.

It is often forgotten that *Frankenstein: or, The Modern Prometheus* begins aboard ship in the Arctic Ocean, as a narrative recounted by a sea captain bound for the North Pole. In a series of letters addressed to his sister in England, Roger Walton conveys ambitions and tribulations that paint him as an archetypal explorer, in whom delusions of grandeur mix with sentimental indulgences of joy and despair. As he faces the trials of leadership and vicissitudes of fortune in the face of nature's extremes, he yearns for a friend to temper his judgments and in which to confide his emotions. One morning, an uneasy crew awakens to find their ship immobilized in the frozen sea, surrounded by an impenetrable fog. There, against all odds, Walton's wish for genteel company finds fulfillment with the rescue of a half-frozen man traveling on foot over the shifting surface of the ice. His determination to reach the pole outmatching even that of the ambitious explorer, Victor Frankenstein inexplicably refuses to board the ship until assured of its destination. Today's reader knows the reason for Frankenstein's absurd pursuit even before he discloses it to the novel's narrator; Victor seeks the death of the monstrous being he has brought to life by artificial means, only to abandon in horror and regret. Having first pleaded for love and sympathy from his virgin father, the nameless creature now seeks vengeance, and their struggle unto death has driven them to the very ends of the earth. The tale would hardly be plausible had the sailors not witnessed an apparition in the fog that corresponded to Victor's description of a human form enormous in stature and impervious to the cold—the harshness of the environment a measure of his monstrosity. The novel thus begins with a pleading for the veracity of the story to follow, a demand that the reader draw inferences from the hazy edges of experience to still more marvelous and terrifying possibilities beyond.

*Frankenstein*'s opening scenes in the Arctic double as a parable of the novel's origin story on the banks of Lake Geneva, amid the stormy weather of the "year without a summer." Both the fictional space of the novel, and the historical circumstances of its conception, attest to how a sense of climatic estrangement opens the mind to previously unfathomable notions. Written amid a climate crisis far more mysterious to Mary Shelley and her contemporaries than anthropogenic climate change is to us today, *Frankenstein*'s text and context hold clues for understanding how attention, perception, and cognition are moderated under atmospheric conditions animated by unseen forces. This moment is exemplary, for unfamiliar things tend as often to be overlooked as to inspire curiosity. Fright may induce repression rather than provoke confrontation. And if strange phenomena do set the imagination to work, it may be in the direction of paranoia as opposed to insight—psychological dynamics clearly at work in the contemporary culture of denial surrounding climate change. Perhaps more insidious than outright rejection of the scientific consensus, aversion of attention offers a passive means of avoidance, allowing problems that warrant an immediate response to slip below the radar even in contexts where climate science is accepted.[2] Such tendencies are often condoned by social norms of attention, which, for example, preserve "talking about the weather" as a banal space of neutral discourse. What makes it possible to notice aberrant phenomena and hold them within our sphere of attention, without freezing in shock or turning away in disbelief?

*A Year Without a Winter* stages a series of environmental disorientations designed to foster dwelling within spaces of uncertainty and ambiguity that typify forecasted climate futures. In contrast to the imperative of activist environmental art, which aims to clarify understanding of climate change and offer a take-home message about what to do about it, this project posits the need for deeper aesthetic explorations of the uncomfortable territory of ambiguity, indeterminacy, and our inextricable

---

**2** Kari Marie Norgaard, *Living in Denial: Climate Change, Emotions and Everyday Life* (Cambridge, MA: MIT Press, 2011); Eviatar Zerubavel, *Hidden in Plain Sight: The Social Structure of Irrelevance* (New York: Oxford University Press, 2015).

complicity in today's entangled ecological crises. Amitav Ghosh has recently argued that climate change remains *unthinkable* within a dominant cultural logic epitomized by the realist novel, the aesthetic conventions of which consign uncanny affects, rare and implausible events to the subgenres of science fiction, fantasy, and horror.[3] Horror is a genre populated by monsters often hailing from netherworlds or outer space, beings whose very existence is a "violation of the natural order, where the perimeter of the natural order is determined by contemporary science."[4] The confluence of Gothic aesthetics with Tambora-era climate valorized being overwhelmed by sublime landscapes and fascinated by monstrous visions, giving rise to new literary genres beyond the pale of realism. From colonial ambitions in the Arctic to aesthetic ambitions in the Alps, and later, lingering cold and famine, multiple interests converged in this historical moment reward attention to climate. And yet, as the monster stories of Shelley and John Polidori attest, it is not necessarily the image *of* climate or nature that sustains the strike of lightning, the torrents of rain, the fog that blankets the mind in mysterious environs. One cannot predict what delusions and insights may arise *within* an unfamiliar environment. Climate makes itself felt not only as manifest content (a preoccupation with ice and darkness) but more subtly, as context for strange affects and impossible beings. An imaginative grasp of climate change demands new ways of toggling between figure and ground, allowing the diaphanous and unstable conditions of climate to arrest attention as fellow protagonists in our contemporary drama. If these conditions evade comprehension, this is due as much to aesthetics of art and nature as to the idea of climate itself.

The concepts of climate, atmosphere, air, and ambience present a slippery constellation of objects and metaphors. Connoting at once the transparent and omnipresent background condition of worldly affairs and a specific set of conditions to be measured and moderated, the influence of atmosphere can be hard to recognize, much less to control. Like tools, which,

---

**3** Amitav Ghosh, *The Great Derangement: Climate Change and the Unthinkable* (Chicago: University of Chicago Press, 2016).

**4** Noel Carroll, *The Philosophy of Horror: or, Paradoxes of the Heart* (New York: Routledge, 1990).

according to the Heideggerian phenomenological tradition, we use without noticing their particular properties, climate calls attention to itself only when it loses its primary attribute of reliability. Bad weather, pollution, smoke conditions, and invisible toxins bring a disturbing hypervisibility to the sky.[5] Geographer Mike Hulme argues that the idea of climate is an abstraction of the human mind and its methods of calculation whose function is to introduce "a sense of stability or normality into what would otherwise be too chaotic and disturbing an experience of the unruly and unpredictable weather."[6] As the object of atmospheric science, "climate" denotes the statistical average of weather conditions at a particular locale ranging in size from a room to the globe, over a period of time, from seasons to millennia; thirty years, by the standards of the World Meteorological Organization. In contrast to the rain on one's skin, climate works upon thought, moderating expectations and near- and long-term strategies for dwelling within varied environs. Architecture, clothing, hunts, and harvests are attenuated to a range of physical affordances of environments, which are rendered (relatively) predictable through cultural ideas of climate. As Cynthia Selin and I have argued elsewhere, climate is a *lived abstraction*, whose contours can be traced in fashion trends and everyday practices of anticipation such as dressing for the weather.[7] As climates change physically and our forms of life within them evolve, for example, through the use of climate control technologies, transportation, communications, and the fuels that power them, cultural ideas of climate must adapt as well if they are to continue to serve their social function on instilling a sense of order within the ever-fluctuating atmosphere. The increasingly variable physical condition of the atmosphere today thus shifts a heavy burden of anticipation onto culture.

Anthropogenic climate change challenges the cultural imaginary on many levels, first and foremost because it flies in the face of the aspiration to stability. "Expect the unexpected"

---

5 Peter Sloterdijk, *Terror from the Air* (Los Angeles: Semiotext(e), 2009).

6 Mike Hulme, "Climate and Its Changes: A Cultural Appraisal," *Geo: Geography and Environment*, vol. 2, no. 1 (May 21, 2015): 1–11, 2.

7 Dehlia Hannah and Cynthia Selin, "Unseasonal Fashion: A Manifesto," in *Climates: Architecture and the Planetary Imaginary*, ed. James Graham et al. (Zürich: Lars Müller Publishers, 2016), 222–231.

is the mantra of workers at the front lines of climate chaos (see Pablo Suarez in this volume), a monstrous violation of the very idea of climate. Yet it is not merely the increased variability of weather but the *scales* of abstraction that challenge our capacity to think climate and its changes today. In their study of the ancient Greek concept of *klima*, for example, James Fleming and Vladimir Jankovic argue that climate has long been conceived as an *agency* affecting human bodies and societies rather than an *index* of physical patterns of the atmosphere, as it is in today's scientific world picture.[8] Historically, ideas of climate were attuned to local geographies and the forms of human and non-human life to which they were imagined to be conducive. In the Aristotelian imaginary, climates were plural, divided between gracious temperate regions and inhospitable torrid and frigid zones. Climatic and geographic extremes converged near the Earth's equator and its poles. In the early nineteenth century, Alexander von Humboldt would add altitude to consideration, defining climate as "*all the changes in the atmosphere which sensibly affect our organs*, as temperature, humidity, variations in the barometrical pressure, the calm state of the air or the action of opposite winds, the amount of electric tension, the purity of the atmosphere or its admixture with more or less noxious gaseous exhalations, and, finally, the degree of ordinary transparency and clearness of the sky... *with reference to the feelings and mental conditions of men*."[9] Any understanding of anthropogenic climate change relies on a concept of global climate that emerged only in the twentieth century, with the development of vast weather monitoring and communications infrastructure that made it possible to track wind patterns, ocean currents, and states of the atmosphere worldwide.[10] The atmospheric sciences treat climate as an index, a set of empirical data to be analyzed and modeled into global statistical abstractions, yet the discourse of agency pervades discussions of the effects of climate change on economies, environmental justice, political conflicts, migration, food security, and

---

**8** James Rodger Fleming and Vladimir Jankovic, "Introduction: Revisiting Klima," *Osiris*, vol. 26, no. 1 (January 1, 2011): 1–15.

**9** Fleming and Jankovic, "Revisiting Klima," 5–6. Emphasis mine.

**10** Paul N. Edwards, *A Vast Machine: Computer Models, Climate Data, and the Politics of Global Warming* (Cambridge, MA: MIT Press, 2013).

so forth. What does it mean to be subject to the ghostly agencies of an abstraction, a global climate unevenly saturated with the effluence of agencies of our own? How and where does one encounter such distributed agencies and their effects?

Today, as the polar vortex whips cold air down to lower latitudes from a warming Arctic, mountains lose their snowcaps, and the Gulf Stream holds hot air stagnant over the Atlantic, climatic extremes betray familiar geographies. Concomitantly, travel, personal communication, and mass media proliferate our exposure to environmental conditions at disparate locations worldwide, mixing in cultural space even as they swerve in physical space. Weather speaks to us from all directions in a polyphonic babble, which resists abstraction to stabilizing patterns. Before climate change can be clarified, the idea of climate must be rethought. And in order to be rethought, it must be reencountered, through aesthetic forms, images, narratives, and vectors of experience through the world itself.

> A *fieldwork* engages with the geographical site but then warps one's perception of the space comparable to a mathematical "strange attractor." Sharing, on the one hand, the history of art installation (which can modulate the encompassing architecture of the viewer's phenomenological perception) and, on the other hand, the history of "site-specific" or earthwork art (which amplifies the place's history or materiality), a *fieldwork* creates its own temporary architecture within a space or in a landscape. However, such a landscape need not be natural, and the architecture may not always be a traditional shelter or sculpture but can be composed of sonic material, electromagnetic fields, light fluctuations, or relationships. At its core, a *fieldwork* is dynamic and geospatial.[11]

Charles Stankievech's single channel video *LOVELAND* captures purple smoke from a military grenade sweeping across an icy

---

**11** Charles Stankievech, *Magnetic Norths: A Constellation of Concepts to Navigate the Exhibition* (Montréal, Qc: Galerie Leonard Bina Ellen Art Gallery, 2010).

Arctic landscape, a gesture that invokes the social and environmental legacies of the Cold War and Canadian militarization of the Arctic, a subject with which the artist has engaged extensively through his fieldworks in the region.[12] Its optical quality has a very different etiology, however. Perhaps it was Stankievech's sensitization to the unique color palette of northern latitudes that enabled him to imagine this landscape as a staging ground for the realization of the aspiration of another artist. The color field painter Jules Olitski once said that he would prefer "nothing but some colours sprayed into the air and staying there." *LOVELAND* is a reconstruction in another medium of Olitski's painting *Instant Loveland* (1968), in which an evanescent lavender mist appears to evaporate off of the indeterminate surface of an enormous canvas, an homage to the painter's ability to capture the exquisite softness and dynamism of tinted air within a static plane. Through Stankievech's remediation of the painting into the dynamic medium of video and sound, the experience of looking at an indeterminate image yields to a synesthetic sense of being engulfed within an illusory atmosphere, a fleeting configuration of the air that subsumes a complex of histories and geographies.

Around the same time that the Villa Diodati group endured mysteriously bad weather at Lake Geneva, Romantic painters such as John Constable, J. M. W. Turner and Caspar David Friedrich developed new ways of capturing the sky. In retrospect, the extraordinary reddish sunsets and storm clouds they depicted were likely effects of industrial pollution, volcanic particulates, and the climatic disturbances caused by invisible gases circulating in the upper atmosphere after Tambora's 1815 eruption. Indeed, paintings of this era have been used by atmospheric chemists to reconstruct the history of the impact of volcanic eruptions on optical aerosol depth—a measure of how the atmosphere absorbs and scatters light and traps heat.[13] Astute visual palpation of that era's strange airs thus corroborates and amplifies phenomenologically the empirical reconstruction of past atmospheric conditions

---

**12** Charles Stankievech, *Loveland* (Berlin: K. Verlag, 2011).

**13** C. S. Zerefos, V. T. Gerogiannis, D. Balis, S. C. Zerefos, and A. Kazantzidis, "Atmospheric Effects of Volcanic Eruptions as seen by Famous Artists and Depicted in their Paintings," *Atmospheric Chemistry and Physics* 7 (2007): 4,027–4,042, https://doi.org/10.5194/acp-7-4027-2007.

based on other forms of proxy data, such as air bubbles and dust trapped in ice cores. The gradual disclosure of latent meaning—in an image or the very air we breathe—in turn recasts the perceptual experience of being lost in a mysterious cloud.

In aesthetic form and methodology, *LOVELAND* assumes the role of an urtext for *A Year Without a Winter*, offering an image that gathers significance with each iteration through different media and narratives. The motif of aesthetic absorption within a colorful mist links Tambora's aftermaths with our present efforts to imagine anthropogenic climate change, yet the attractive force of this image extends still further. Upon leaving the video installation, my glance falls upon a stone endlessly reflected in a mirrored vitrine, glowing magenta under black light—a unique property which identifies the stone as a Chatham Emerald, a synthetic emerald first grown in the laboratory in 1935. Next to it, an old paperback is pinned open to a paragraph that describes an ominous purple cloud. With the lightness with which Shelley recalls that "some volumes of ghost stories... fell into our hands," Stankievech recounts that, two years *after* filming *LOVELAND*, he happened upon a 1901 science fiction novel by M. P. Shiel, titled *The Purple Cloud.*[14] The novel tells the story of a toxic purple plume, emitted by a volcano somewhere in the South Pacific, that envelops the earth and poisons every living creature in its path, leaving in its wake a world strewn with glittering jewels. A sole witness to the catastrophe is spared, having traveled deep into the Arctic in hot pursuit of a prize offered to the first explorer to reach the North Pole. The stone displayed in the vitrine is presented as a dubious trophy collected on his path as he traverses a desolate earth in search of other survivors, an archetypal journey for the protagonist of a work of genre fiction—of which the first instance is Mary Shelley's post-apocalyptic novel *The Last Man* (1826).

*A Year Without a Winter* pursues moments of serendipity and intertextuality between historical and fictional figures, across real and symbolic geographies, artistic media and genres, in search of sites at which abstractions suddenly become palpable, and monsters come alive before our eyes. The project is informed, in this regard, by a sustained meditation on the conditions that

**14** Private conversation with Charles Stankievech, July 10, 2015.

inspired Shelley, who "through her integration of autobiography and allegory, memory and myth, experienced and imagined... conjures up a literary vision that is at once simple and sublime," via which "the mythical is rendered personal and the personal is rendered mythical."[15] In Shelley's life and fictions, a preponderance of such transformative instances occur at nature's rugged extremes. If the poles marked hypothetical limit cases, mountain peaks offered more proximal sites at which to test mind and spirit in contemplation of the order of nature and its implicit moral norms. In *Frankenstein,* geographical limits become staging grounds for Victor's confrontation with his transgression of those norms—transgressions often performed within the interior confines of the laboratory, the house, or indeed the bedroom, where Shelley, in a waking dream, first envisions a hideous creature peering into her window, illuminated by a bolt of lightning.

Victor, the experimentalist, and Walton, the explorer, personify two creative-epistemological modes that recur in the present project. Yet while reception of *Frankenstein* has tended to privilege the novel's experimental (particularly its biotechnological) implications, this project prioritizes the exploration of unfamiliar environs—a reversal reflected in the book's title. Rereading *Frankenstein* and the Year Without a Summer with the hindsight knowledge of the Tambora eruption transforms the geography and periodicity of the historical episode for which Shelley's experience has become a powerful synecdoche. A year stretches over three; a finite period of climate crisis expands into an indeterminate future. Meanwhile, the site of the crisis expands beyond the Swiss Alps, across the northern hemisphere, to Indonesia, and eventually is understood to have affected the whole world, as does climate change today. Within the expanded field of a global climate catastrophe, the logic of the laboratory becomes operative in the world itself; the monster, now in the form of a remixed atmosphere, escapes into the field to live a life of its own. In order to trace its steps physically and imaginatively, we must perform fieldwork, sometimes in the artistic form of fieldworks.

---

**15** Brittany Reid, *From Prometheus to Presumption: Frankenstein's Theatrical Doppelgänger* (Master's Thesis, Department of English, Calgary, Alberta, 2013), 42. Quoted with permission of the author.

**Jules Olitski, *Instant Loveland*, 1968. Copyright Estate of Jules Olitski. Licensed by VAGA, New York, NY. ©Tate, London 2017.**

**Charles Stankievech, *LOVELAND*, 2009–2011. Interior view of mirror vitrine with circa 1930 Chatham Emerald Cluster fluorescing under Ultraviolet Light. Courtesy of the artist.**

With childlike fascination, Victor recalls a tree felled by lightning during a thunderstorm, a dazzling sight that opens up a fateful turn in his intellectual development. In a *mise en abyme* of Shelley's confinement to the Villa Diodati, Victor is introduced to natural philosophy during an Alpine spa holiday, when "the inclemency of the weather obliged us to remain a day confined to the inn." "In this house," he recalls, "I chanced to find a volume of the works of Cornelius Agrippa." While his childhood companions play outdoors, young Victor becomes obsessed with the inner workings of nature and retreats into the library to study the archaic alchemical texts. Dreaming of transmuting other metals into gold and discovering the philosopher's stone, he is impressed by a lecture on the new theory of electricity. Galvanism seems to offer a far superior explanation of the tree's destruction, proving key to his tragic discovery of the secret of life. Realizing that "the modern masters... can command the thunders of heaven, mimic the earthquake, and even mock the invisible world with its own shadows," Victor relinquishes alchemy and submits himself to instruction in the new science of chemistry. In words that closely paraphrase those of Sir Francis Bacon (the notorious father of experimental science whose masculine failings offer a clear archetype for Victor), he follows the moderns in seeking to "penetrate into the recesses of nature and show how she works in her hiding-places."[16] For Bacon, the aim of art (*techné*) is to imitate nature and bring its powers under artificial control, to "follow and as it were, hound nature in her wanderings" so as to "be able to lead her to the same place again."[17] From his observation of nature at work in an act of destruction, Victor is inspired to follow nature's wanderings into the secret recesses of creation, a methodology that Shelley extrapolates just one step beyond reality into science fiction.

In his tumult of emotion, Victor is portrayed as a modern experimentalist in the dangerous grip of archaic sentiments. Wonder, as historians of science Lorraine Daston and Katherine Park have shown, was granted pride of place in the psychology of European natural philosophical inquiry from at least the medieval period until the scientific revolution.[18] Only by the mid-eighteenth century would the dispassionate scientific persona that we are familiar with today emerge as a norm. Awe, reverence, and

---

**16** Shelley, *Frankenstein*, 76.

**17** Indeed the tradition of feminist critique of Bacon trades upon Shelley's more familiar characterization. Francis Bacon, "The Advancement of Learning" (1605) in *The Works of Francis Bacon: Vol. 3*, ed. James Spedding et al. (London: Longman, 1878), 331.

**18** Lorraine Daston and Katherine Park, *Wonders and the Order of Nature, 1150–1750* (New York: Zone Books, 1998).

fascination with the ways of nature were not merely permissible expressions; these emotions, subsequently banished to the literary domain, were considered epistemic virtues instrumental to the discovery of nature's most intimate and mysterious ways. Wherever something appeared marvelous—a strange birth, a stone that glowed in the dark, a divine face in the clouds, a mermaid, a unicorn, or a platypus—there lay an invitation to comprehend the rare and exotic ways that nature could manifest its possibilities. Or its limits. As one fourteenth-century monk put it, "At the farthest reaches of the world often occur new marvels and wonders, as though Nature plays with greater freedom secretly at the edges of the world than she does openly and nearer us in the middle of it."[19] The boundary between the natural and the supernatural was marked by the hazy zone of the preternatural—sometimes just the unfamiliar. Within this cosmology, one learns what is possible in this world by traveling to its edges, where nature plays more freely, if also under greater constraints. Only in the wild and mysterious polar regions could Victor persuade Walton of his mastery of the powers of nature. If Victor's crime is his transcendence and transgression against the natural order, it must be understood as a crime of passion. Wonder leads inquirers to the edges of the world, where monsters are not only discovered but created.

If nature plays more freely at its edges, its center is typified by regularities with which it is the business of science to make us familiar. Nature's geographic and epistemic periphery is thereby brought within the familiar domain of the middle, where the norm is the eternal return of the same.[20] Whether we follow nature deep into the laboratory or to the edges of the world, the ability

**19** Daston and Park, *Wonders and the Order of Nature*, 25.

**20** The idea of nature as a stable background for human history is classically stated by Hegel: "The changes that take place in Nature—how infinitely manifold soever they may be—exhibit only a perpetually self-repeating cycle; in Nature there happens 'nothing new under the sun'...only in those changes which take place in the region of Spirit does anything new arise." Georg Wilhelm Friedrich Hegel, *Introduction to the Philosophy of History*, trans. Leo Rauch (Indianapolis: Hackett Publishing Company Inc.,1988),56. As Dipesh Chakrabarty argues, the division between natural human history is called into question by anthropogenic climate change and other facets of the Anthropocene. Dipesh Chakrabarty, "The Climate of History: Four Theses," *Critical Inquiry*, vol. 35, no. 2 (2009): 197–222.

to reproduce our findings, to "lead her to the same place again" is the foundation of scientific authority. Just as the reports of travelers want for corroboration, experiments must be able to be repeated. For those who cannot bear eyewitness to a phenomenon, trust is garnered through narration, documentation, and assessment of its plausibility.[21] There are conventions for producing trust in both scientific and literary genres, and today's dry scientific article is the outcome of stylistic evolution from ancestors that are hardly recognizable as such, including the alchemical texts that first entranced Victor. Indeed, a debate has raged on as to whether they conducted experiments at all or whether their writings were mere chimeras better fit for analysis on the psychoanalytic couch than in the laboratory. Historian Lawrence Principe has sought to demonstrate, contra an influential interpretation advanced by Carl Jung, that a discontinuity of rhetorical style masks a deeper continuity of material practice between alchemy and modern chemistry. One can be forgiven for reading alchemical texts, with their "frequent references to hermaphrodites, flowers, dragons, kings, queens, and a multifarious menagerie of real and mythical creatures involved in everything from birth and marriage to incest and death" as poetic descriptions of unconscious dramas rather than a protocol for chemical synthesis.[22] Yet, Principe contends, if alchemical texts are obscure, it is because they were meant to guard ritual and trade secrets. In order to prove this, he attempted, like Victor, to operationalize their arcane instructions, focusing on the fabled "seed of gold," or "philosopher's tree," often shown as a phallus sprouting from the body of a dead man, which appears in one fifteenth-century illustration, as a "tree enclosed within a small circular edifice that is almost too small for such a plant."[23] After a laborious effort of literary interpretation and chemical experimentation, one can imagine Principe's delight upon arriving in his laboratory one day

---

**21** Frederic L. Holmes, "Scientific Writing and Scientific Discovery," *Isis*, vol. 78, no. 2 (June 1987): 220–235; Steven Shapin and Simon Schaffer, *Leviathan and the Air-Pump: Hobbes, Boyle, and the Experimental Life* (Princeton, NJ: Princeton University Press, 1989).

**22** Lawrence M. Principe, "Apparatus and Reproducibility in Alchemy," in *Instruments and Experimentation in the History of Chemistry*, ed. Frederic L. Holmes and Trevor Levere (Cambridge, MA: MIT Press, 2000), 55–74; 55.

**23** Principe, "Apparatus and Reproducibility in Alchemy," 55, 67.

to discover a crystalline form tightly fitted inside the bulb of a test tube looking, for all the world, like a miniature metallic tree. (And only on this basis can I expect the reader to follow me in being moved by the sight of a frozen plant inside a vitrine, which now graces the cover of this book...)

By translating the messages latent in this textual menagerie, Principe not only corroborates a hypothesis but also demonstrates a method of historical inquiry. By means of reconstructing past events, one achieves a new understanding of what is possible *to know* and *to do* in this world, redrawing the boundary between fantasy and reality. In a surprising passage in the *Critique of Pure Reason*, a philosophical study on the limits of knowledge, Kant proffers an evocative image of the mind itself as an island shrouded in mist and surrounded by seductive seas.

> We have now not merely explored the territory of pure understanding, and carefully surveyed every part of it, but have also measured its extent, as assigned to everything its rightful place. This domain is an island, enclosed by nature itself within unalterable limits. It is the land of truth—seductive name!—surrounded by a wide and stormy ocean, the native home of illusion, where many a fog bank and many a swiftly melting iceberg give the deceptive appearance of farther shores, deluding the adventurous seafarer ever anew with empty hopes, and engaging him in enterprises which he can never abandon and yet is unable to carry to completion.[24]

Through this analogy, the philosophical imperative to survey the island—to correctly discern what can be known from its highest vantage point, who inhabits it, and by what right we can rest upon it—echoes through Shelley's narrator, as Walton, on his endeavor to reach the North Pole, finds himself instead plumbing the depths of the human soul. A thick fog doubles readily as a projection screen for latent desires and unchecked assumptions. And yet, where the atmosphere itself is the problem, there is no use waiting for the mist to clear away. Where the horizon

24 Immanuel Kant, *Critique of Pure Reason*, ed. Howard Caygill, G. Banham, and Norman Kemp Smith, second edition (Basingstoke: Palgrave Macmillan, 2007).

Julian Charrière, *Tropisme* (2014), plant sheathed in ice inside a refrigerated vitrine. Image courtesy of the author.

An alchemical logic transcending the categories of animal, mineral, and vegetable connects the tree struck by lightning (which inspires Victor in the art of destruction and creation), with the synthetic emerald exhibited by Stankievich, the Philosopher's Tree reconstructed by Principle, the frozen plant preserved by Charrière, the jungles burned to make way for oil palm plantations addressed in *Postcards from Tambora—An Invitation to Disappear*, and Karolina Sobecka's clouds reconstructed in glass jars in the following chapter, "A Memory, An Ideal, and a Proposition."

The Philosopher's Tree, grown inside of a glass flask from the "seed of gold" and distilled mercury. Originally published in Lawrence Principe's "Apparatus and Reproducibility in Alchemy," in *Instruments and Experimentation in the History of Chemistry*, eds. Frederic L. Holmes and Trevor Levere (Cambridge, MA: MIT Press, 2000). Courtesy of Lawrence Principe.

is obscured, one ought not only to peer more intently into the mists but also seek surer footing on the ground beneath one's feet. Sometimes, the only way to do so is to retrace our steps, reexamining along the way the assumptions, methods, and ideas through which we encounter our uncertain environs.

Turning from experimentation to exploration, another striking example of reenactment as a mode of historical investigation is offered the 1947 Kon-Tiki expedition, led by anthropologist Thor Heyerdahl, which endeavored to prove that the Polynesian islands of the South Pacific were originally peopled by ancestors of the Incas from modern-day Peru (see image p. 302).[25] In the symbolic realm, a striking congruence between legends, linguistic elements, artifacts, and seafaring techniques offered ample support for his theory; the only objection was that it would have been physically impossible to cross the Pacific Ocean on a handmade balsa wood raft of the sort used centuries ago. Aboard a meticulously reconstructed vessel, Heyerdahl and a crew of seven managed to do just that, paving the way for the further articulation and acceptance of his geographical theory. Like an experiment that can be repeated in the laboratory, the Kon-Tiki expedition and the historical migration it purports to reenact rely on the stability of natural patterns themselves. "The distance is not the determining factor in the case of oceanic migrations, but whether the wind and the current have the same general course, day and night—all year round," Heyerdahl concludes. "The trade winds and the Equatorial Currents are turned westwards by the rotation of the earth, and this rotation has never changed, in the history of mankind."[26]

While the rotation of the earth has not changed, nor have the laws of physics, the winds and ocean currents are indeed changing under anthropogenic pressures. The perturbation of natural patterns throws into fundamental question the adequacy of both our narratives of nature and our methods of investigation to the worlds we inhabit today—and tomorrow. The consequences are felt everywhere from the dissonance between traditional agricultural practices and changing growing seasons to the

---

**25** Thor Heyerdahl, *The Kon-Tiki Expedition: By Raft across the Pacific*, trans. F. H. Lyon (New York: Rand McNally, 1951).

**26** Heyerdahl, *The Kon-Tiki Expedition*, 230.

cultural unconscious that refuses to acknowledge an environmental crisis that is unfolding right before our eyes. What is needed within this state of epistemic disorientation is a capacity for sustained attention and critical reflection that may give rise to new ways of orienting ourselves within a rapidly changing global climate. Inspired by the Villa Diodati group's literary mediation of the Tambora crisis and the science of their day, *A Year Without a Winter* seeks such epistemic and aesthetic dispositions in contemporary art and literature.

As a method of historical investigation, as in popular cultural practices of historical reenactment (of battles, for instance), the paramount aim is to achieve fidelity to the past. In art, reenactment is often a form of reinvention, governed less by standards of accuracy than an imperative to transmute old stories into new ones, adapt them to different contexts, or subvert previous orders of meaning. Consider Shezad Dawood's film *Feature: Reconstruction* (2008), a postcolonial mashup of cowboys-and-Indians movie tropes, in which Krishna and Crazy Horse tangle with queer cowboys and Wagnerian Valkyrie in a collage of Potemkin villages.[27] In contrast to the previous examples, which sought to unravel mythologies and images in the service of a more accurate account of past events, Dawood's film introduces mutations into historical fact and fiction, producing a new and indeterminate narrative that retains its familiarity through deliberate inversions and substitutions of genders, Indians, real and fictive geographies, Westerns and Easterns. The intelligibility of the work relies on the viewer's capacity to project familiar characters into unfamiliar situations, and conversely, to read displaced figures into familiar landscapes. If we are to better imagine ourselves within the increasingly unruly world described by climate models—a world that itself betrays the norms of realism—the interpretive capacities cultivated both by faithful and parodic practices of reenactment will be crucial to tracking the permutations of familiar cultural ideas of climate, and of ourselves as actors on its stage.

Completed the same year that *Frankenstein* was published, Caspar David Friedrich's *Wanderer above the Sea of Fog* (1818)

27 Shezad Dawood, *Feature: Reconstruction* (London: Ram Distribution, 2008).

offers a paradigmatic representation of the sublime Alpine vistas sought by travelers on the Grand Tour. An image of masculine triumph and reverence for nature's grandeur, the painting depicts a gentleman from behind, standing atop a rocky mountain peak, surveying the range beyond and clouds below. Echoing the philosophical tradition of Edmund Burke and Immanuel Kant, the capacity of the sublime to fortify the mind and elevate the soul are valorized repeatedly by Shelley's protagonists. On the dangerous ascent of Mont Blanc, Victor recalls, "the effect that the view of the tremendous and ever-moving glacier had produced upon my mind... [it] filled me with a sublime ecstasy that gave wings to the soul and allowed it to soar from the obscure world to light and joy." Like Friederich's lone wanderer, Victor continues, "I determined to go without a guide, for... the presence of another would destroy the solitary grandeur of the scene."[28] At its aesthetic extremes, as much as its climatic and geographical limits, nature calms and releases the imagination, at the same time opening it to confrontation with new ideas. Atop the glacier, Victor finds a fleeting oasis from his monumental guilt in the form of temporary amnesia of his actions. Exactly in this moment of oblivion, the creature appears to confront him for the first time, bounding up the mountain at superhuman speed to demand his filial due. In the figure of Victor, the *Wanderer above the Sea of Fog* is drawn down from his mountain pedestal and into the mists that swirl ominously around the world below.

Today we find ourselves within a troubled atmosphere, a place which affords no distant vantage point, bird's eye view, or comfortable illusion of innocence. One must find places for the imagination to play within disturbed concentrations of atmospheric gases, clouds of smoke, burning forests, and commodified skies. These conditions offer a constant reminder that we do not wander alone. From aesthetic encounters with contemporary environments there must emerge new cultural ideas of climate, notions which nonetheless retain their familiarity and continuity with older senses of the term through deliberate conceptual inversions and category violations. We must develop an idea of climate as figure rather than ground, which highlights the range

**28** Shelley, *Frankenstein*, 123.

**Caspar David Friedrich, *Der Wanderer über dem Nebelmeer* [Wanderer above the Sea of Fog], ca. 1818.**

of variation in weather rather than the mean, takes account of the reciprocal influence of human activities on "all of the changes of the atmosphere," comprehends the relation between scales of abstraction from local to global, and tracks associations between short-term variations and long-term trends—a domain in which we are irreducibly imbricated, complicit, and far from disinterested. Ultimately, it must render the monstrous intelligible within the domain of the real. If a striving for stability is a central feature of the idea of climate, then perhaps this aspiration will have to be located less in the dynamics of the atmosphere than in our means of mediating changing environments—through model predictions, short-term weather forecasting, architecture, apparel, and everyday habits adapted to abrupt climatic shifts.

Although the *Wanderer above the Sea of Fog* is taken to portray the landscape at a distance, in retrospect we can infer that Tambora's efflux suffuses the air, filters the light, and blocks out the sun. Unbeknownst to Friedrich, his hero wanders as much *within* as *above* this portentous mist. Tracing this cloud back to its historic source with the Swiss artist Julian Charrière, the trope of mountain sublimity was superseded by other environmental logics and aesthetic ideas. As we descended from Tambora's summit through an ethereal fog, we traversed clouds emanating from trash fires, airplane engines, and burning jungles. Further below, and on nearby islands, the mountain's sparse vegetation gave way to endless plantations of oil palms and rubber trees. Global demand for palm oil fuels forest fires that burn annually during the late-summer dry season, blanketing a vast area of the Asia Pacific region in a toxic haze, destroying animal habitats and important carbon sinks. And yet, like the extreme environments that entranced Walton and the Romantics, these complex, anthropogenic scenes also afford inspiration—a story told in this volume's *An Invitation to Disappear—Postcards from Tambora.* We went to Tambora to visit one site, but we encountered another, a landscape to which we relate through a complex network of economic, geophysical, and cultural connections. More often, we are disconnected from the environmental effects of an anonymous ingredient in processed foods, cosmetics, and biofuels. Palm oil comes in jar, too. Among those who work and live within the environs that supply this golden liquid, other suggestive ontologies arise. As anthropologist Sophic Chao's research reveals, the Marind indigenous communities of West Papau encounter the reciprocal problems of deforestation and the expansion of oil palm plantations through a multi-species cosmology in which trees are construed as important beings, or persons, within an "ecology of selves."[29] As new arrivals within this cultural ecology, oil palms are widely construed as problematic community members, threatening the well-being of others, behaving in uncooperative and incomprehensible ways, and therefore defying relations of kinship between individuals of many species within the forests that they displace. Some of Chao's informants even pity the trees for their alienation from loving community, and speculate that oil palms may be unwitting agents through which other agendas

**29** Sophie Chao, "In the Shadow of the Palm: Dispersed Ontologies among Marind, West Papua," *Cultural Anthropology*, vol. 33, no. 4 (November 2018): 621–649.

of environmental violence are enacted. Like Shelley's creature, they are monsters unleashed upon the world by mysterious agencies, personifications of a global community's absent presence within a changing environment.

*An Invitation to Disappear* (2018) is a filmic expedition into the heart of a lush dystopian landscape symptomatic of the current global derangement of ecological thinking.[30] Set in an oil palm plantation at an unmarked location, a nowhere that could be anywhere, in Indonesia or Malaysia—where almost 90 percent of the world's palm oil is now produced—the film stages a disturbing confrontation with the conflicting promises of two global monocultures: rave culture and industrial agriculture. Entranced by a vague sense of promise, the camera slowly traverses a turbulent cloud, which gives way to reveal row upon row of oil palms, heavily laden with fruit, and spreading out in every direction. As the waning light of dusk penetrates the forest's thick canopy, the grid cast on the ground by the sun's last flickering rays is replaced by flashes of light deep in the distance. Blending with the sounds of the forest, a low rhythmic techno beat is felt before it is heard, inducing a sense of direction within the nauseating infinity of the grid. Drawn steadily toward a dark mirage by the rising sound, the camera moves slowly through the darkness until it happens upon a scene of jubilation. An enormous sound system is illuminated between flashes of strobe light. In the conspicuous absence of people, a party rages with mesmerizing intensity. Riveted by the scene, the camera moves slowly straight through the row of palms in an interminable shot, broken only by the smothering effects of gusts from seemingly autonomous smoke machines. As the night wears on, delirium sets in; the base pounds relentlessly upon the deaf ears of the monoculture planation, endlessly deferring an implied climax of collective consciousness.The dim light of dawn finally cuts through the palms fronds, blinding the camera momentarily in a swirling mist, through which the expedition resumes in an infinite loop.

**Following: Julian Charrière, *An Invitation to Disappear*, 2018. Courtesy of the artist.**

---

**30** For an extended discussion of the exhibition, see Dehlia Hannah, "Field Philosophy," in Julian Charrière, *An Invitation to Disappear* (Roma Publications, Musée des Bagnes & Kunsthalle Mainz, 2018).

## A Memory, an Ideal, and a Proposition

Karolina Sobecka

In this project, three clouds that changed the world are reassembled. The material composition of the original clouds is reconstituted, and the conditions necessary for the cloud formation are applied.

Clouds created for this project are models of a cloud that formed in the past (a memory), a cloud that formed in a lab (an ideal), and a cloud that is proposed to be created (a proposition). These particular clouds transformed how we think about climate, technology, and human command of nature. By examining their material composition and the conditions in which they formed, the project aims to rethink the reality of geological and social transformations they paved the way for.

Clouds acquire characteristics of the ground below them. Particles at the center of cloud droplets are traces of natural and human activity. Thus every cloud is a material memory of one unique assembly of what David Gissen calls "socio-nature." To paraphrase Gissen, each cloud is more than a sum of gasses, matter, and forces: it "contains within it the tragedies and successes of the social transformation of nature that exist wherever human experience appears."

◆

## CLOUD A
(A MEMORY)

**Date:** April 5, 1815

**Location:** 2–20 miles above the Mount Tambora volcano on the island of Sumbawa, Indonesia. Having reached the stratosphere, the volcanic cloud first covered the local region and then spread out along the equator, eventually reaching the poles and veiling the globe for several years.

**Chemical composition:** The cloud condensation nuclei in Cloud A is pulverized rock emitted from Mount Tambora during the volcano's eruption. The rocks are black, glassy, biotite-bearing netrachyandesites—a highly unusual type. 42 cubic miles of this rock were emitted into the atmosphere, along with 55 million tons of sulfur-dioxide gas, which combined with hydroxide gas in the stratosphere to form sulfuric acid that condensed into tiny droplets.

**Conditions:** The concentration of atmospheric CO2 was approximately 280 parts per million (ppm), what is today considered a pre-industrial level. It was also in 1815, at the brink of the Industrial Revolution, that coal output soared: Britain alone produced 23 million tons.

**Effects:** The immediate effects of the cloud were widespread and lasted several years. It created an agricultural disaster, with food riots and epidemics breaking out throughout Europe. Red and brown snow fell throughout the year in Europe and Asia. The weather presented "the appearance of vast and dreadful desolation" (as Mary Shelley notes in her journal) and inspired many expressions of anxiety and dread, including the literary creations of Frankenstein and Dracula.[1] The use of carmine by artists increased, a reflection of scarlet sunsets. A century and a half later, scientists discovered that Cloud A caused a climate anomaly, reflecting sunlight and cooling the globe by 0.7–1.3°F.

---

1 Julian Marshall, *The Life and Letters of Mary Wollstonecraft Shelley*, vol. 1 (London: Richard Bentley and Son, 1889), 145.

Cloud A. Courtesy of Karolina Sobecka.

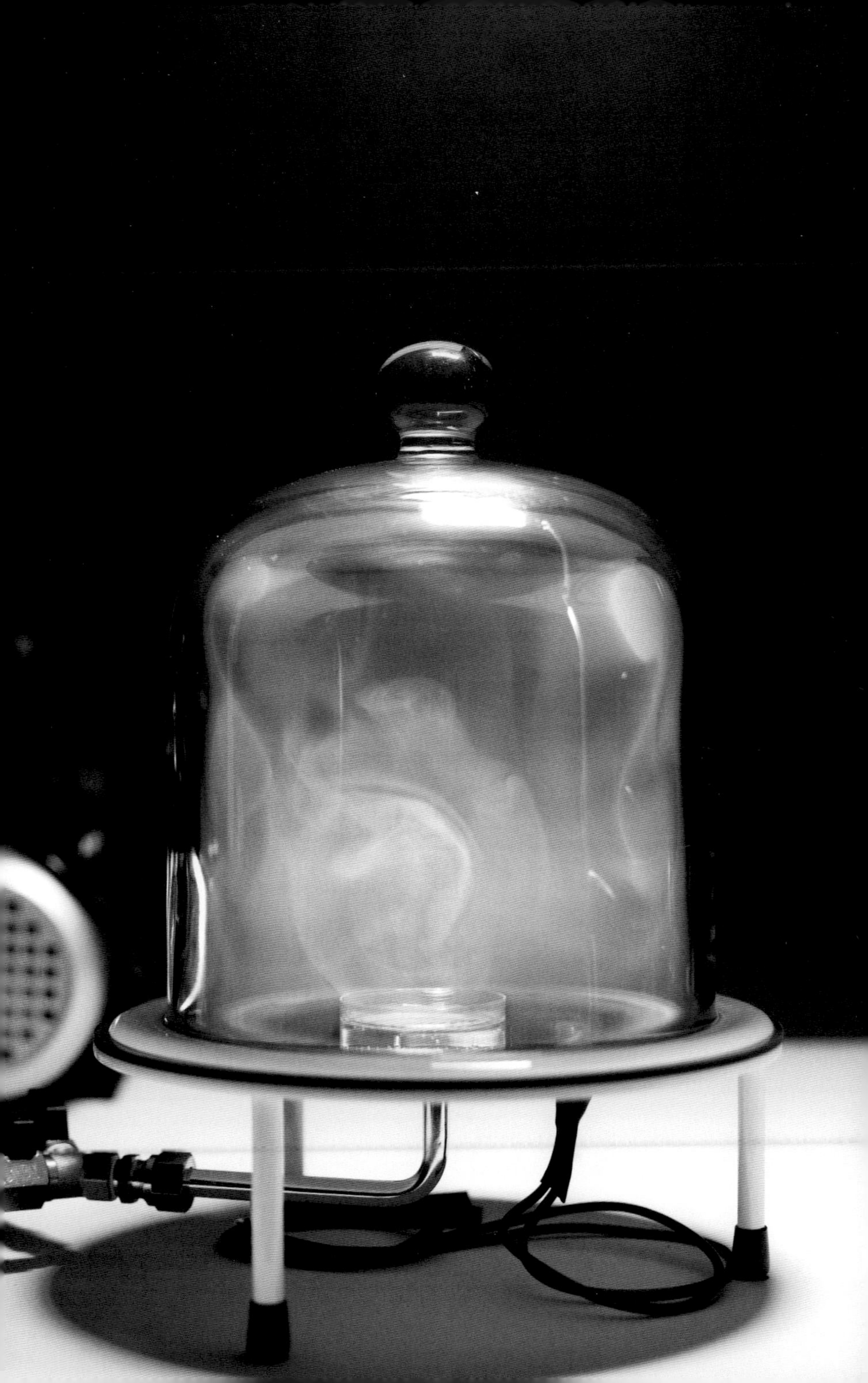

Cloud B. Courtesy of Karolina Sobecka.

◆

## CLOUD B
(AN IDEAL)

**Date:** 1946

**Location:** A cloud formed in a "cold box" at the General Electric Research Lab in Schenectady, New York. The GE cold box became a cornerstone of cloud research, eventually turning into "Project Cirrus," a weather modification program conducted in collaboration with the Office of Naval Research and the Air Force.

**Chemical composition:** 100 grams of dry ice seeded the first cloud created in the GE lab. "Ice-nucleating" particles, such as dry ice, induce water to condense into cloud droplets and freeze at higher than normal temperatures, producing precipitation. On November 13, 1946, Dr. Vincent Schaefer and Dr. Bernard Vonnegut (the brother of novelist Kurt Vonnegut) successfully induced rain in a cloud—"an unsuspecting cloud over the Adirondacks"—and the results were dramatic enough to warrant the creation of a program for modifying the weather. The Cloud B model uses the bacteria *Pseudomonas syringae*, biological ice nucleation particles whose role in cloud formation has only recently been discovered.

**Conditions:** In 1946, the atmospheric concentration of CO2 was approximately 309 ppm. In the postwar atmosphere of techno-optimism, the US was entering a period marked by unprecedented economic growth and new human capability to wield atomic energy—technology powerful enough to have an impact on the entire planet.

**Effects:** Cloud B captured the excitement of the moment, promising to control the weather—and to fulfill the dream of meteorological researchers and military leaders alike. The full command of natural resources through technological means seemed to be just around the corner. But the clouds outside of the cold box proved impossible to master. Within a few years, the discovery of chaos theory described a fundamental limit in the prediction and control of natural systems.

Cloud C. Courtesy of Karolina Sobecka.

◆

## CLOUD C
(A PROPOSITION)

**Date:** 2018

**Location:** 8 miles above the ground, Tucson, Arizona

**Chemical composition:** As part of the first field test in Solar Radiation Management (SRM) research, 100 grams of calcium carbonate will be sprayed into the stratosphere. SRM proposals aim to counteract global warming by shielding the earth from the sun by creating aerosol clouds in the stratosphere. The project was green-lighted in November 2016 and will be conducted in the next eighteen months, after years of controversy. Different particles will be tested during the research, including diamond dust and sulfates, which mimic material emitted during volcanic eruption.

**Conditions:** By 2018, atmospheric concentration of CO2 will be approximately 410 ppm, a dramatic increase from pre-industrial levels, and evidence of human disruption of the climate system. The climate crisis, and the failure to address it over the preceding decades, has triggered the turn to emergency measures exemplified by this geoengineering test. The position of the scientific community and political leadership has slowly shifted—giving serious consideration to climate engineering proposals, despite the enormous controversy around them.

**Effects:** The effects of geoengineering research are difficult to predict. The physical impact on the atmosphere is expected to be entirely benign. But it is the social impacts of going down this path that are a cause for concern and debate.

## Narrating Arcosanti, 1970

James Graham

Architecture has a way of being there. It is durable—in its material forms, its mythologies, its endless circulation of images—while remaining equally mutable. Buildings are never static objects, nor are their contexts, whether spatial, environmental, or social. This twinning of durability and mutability is what makes architecture a fruitful site and subject of narrative, and what makes the field of architecture itself a narrative practice. Buildings never behave quite like their designers design them to (as with the unruly modernist tower of J. G. Ballard's *High-Rise*). Despite the desire of architects to see buildings as "invaluable devices of collective synchronization," to borrow a phrase from Mark Wigley—as objects that locate us meaningfully in time—their time is immediately out of joint and their meanings are immediately unstable.[1] It's not just that the world changes around them and we see architecture anew, as so many objects against an animated backdrop of history and life. Perhaps it's more that an encounter with architecture is something like the confluence of two tidal systems, a liquid friction in which building and world rearrange themselves, an ongoing process that a visitor only catches the briefest glimpse of. That encounter is shaped by the ways we *tell* architecture.

This essay is a narrative about architectural narrative, about architecture as a generator of new narratives, and it takes as its subject a collection of buildings that is often described as a kind of fiction, despite its being here in the world—the "desert

1 J. G. Ballard, *High-Rise* (London: Jonathan Cape, 1975); Mark Wigley, "The Architectural Cult of Synchonization," *October* 94 (Autumn 2000): 37.

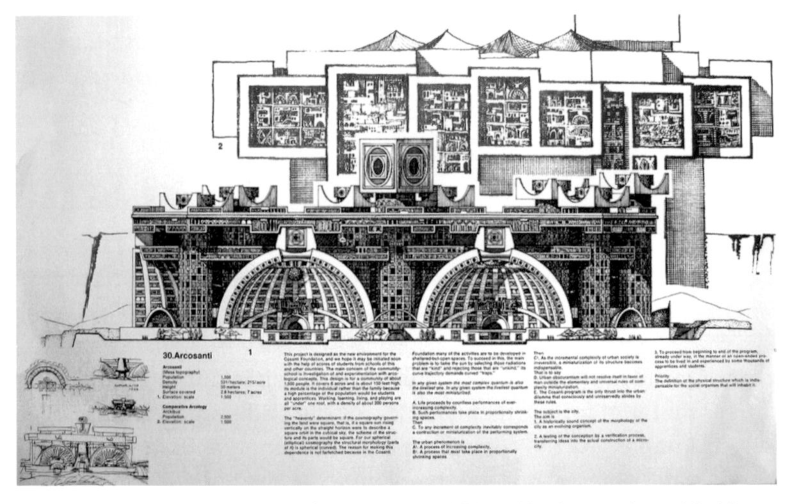


**The original design of Arcosanti (shown with additional background graphics) in Paolo Soleri's *Arcology: The City in the Image of Man* (Cambridge, MA: MIT Press, 1969). This iteration lists a projected population of 1,500. Courtesy of the Cosanti Foundation.**

city" of Arcosanti on the mesas north of Phoenix. Instigated by the architect Paolo Soleri as the 1960s turned to the 1970s, Arcosanti is both building and story, one that congealed into something of an official narrative early on—architecture and architects always exist in a cycle of mythology, with people and places wrapped together in the repetition, reiteration, and revision of such narratives. Arcosanti's story is bound up in Soleri's story, a story that emerges fully only when told in multiple, a story that has long been in need of other voices to participate in the telling. Such voices have emerged in recent months, revealing a darker and long-unacknowledged history; such voices can also be found in Arcosanti's archives, bringing different protagonists and different energies into that history. And further voices can be cultivated, in that objects of architectural fascination like Arcosanti are continually retold and reinterpreted, reflecting the urgencies and imaginations of those who experience them as much as those who ostensibly author them.

Designed in the heyday of the "megastructure" as an object of architectural speculation, Arcosanti was initially envisioned as a new model of the city—a layered urbanism guided by an invented discipline Soleri termed "arcology," architecture meeting ecology. It was an urbanism of three dimensions, pointedly opposing the comparatively flat carpet of infrastructure laid down to accommodate the expansion of US cities in the aftermath of the Second World War, and it was to be colossal in scale, weaving all the constituent elements of city life into a unified design. As drawn in Soleri's *Arcology: The City in the Image of Man*, Arcosanti was to be a 164-foot-high demonstration of these architectural, material, and social precepts. As built over the 1970s, though, its scale and character resembles that of a village, handmade by medium-skilled labor in site-cast concrete. Where the drawings project a tightly ordered aggregation of half-domes and massive concrete plinths, it was realized as a loose assemblage of vaults, half-domed apses, and cubes with circular cutouts, a testing ground of Soleri's characteristic devices. The Arcosanti one visits today speaks to a pragmatically manageable process of communal construction more than the singular urbanity of his arcological drawings.

That Soleri's arcologies are often received as a kind of science fiction, a "paper architecture" that remained in the realm of fantasy, was in part encouraged by Soleri's own exhortations. His interest in framing architecture as something of a space-age technology of vertical mobility explains why he would deliver what was surely a rousing, if occasionally mystifying, keynote to the National Association of Elevator Contractors in 1971. "You, the members of the vertical transportation industry, and I are interested in the same thing, although by different aims," he begins his lecture, titled "The Flight from Flatness." "Gravity sees to it that a leash will always be upon us (unless we can join the Apollo astronauts) but it is up to us to make it more and more elastic."[2] With rhetoric like this—an urbanism that even begins to approach the soon-to-conclude Apollo program—it is little wonder that Soleri's sketches strike some as cover art to a mass-market paperback about extraplanetary life.

His lecture continues by conjuring a grand showdown between the elevator as a heroic figure of urbanity and the true urban Goliath of the twentieth century, the automobile: "The

---

**2** Paolo Soleri, "The Future: The Flight from Flatness," *Elevator World* (1971): 31.

battle will be waged between the 'horizontalists' and their hordes of intruding machines, and the 'verticalists' and their (at present) slim battalions of self-effacing riding boxes and ramps." With a subtle sleight of hand, Soleri upends the question of just whose vision of the city is science fiction and whose is humane realism, portraying an American Dream of infinite automobility and ever faster, ever more extensive interurban transit as a greater fantasy than his own immodest propositions:

> Velocity is contradictory within the urban diaspora.
> It pre-supposes science fictions like instant acceleration and deceleration. It pre-supposes the submission of people, including children, the ill, the aged, etc. to many G's of pressure. It pre-supposes incredibly sophisticated and grimly expensive hardware. It pre-supposes utopia in its most fraudolent [sic] schemes. This is where you "verticalists" are in agreement with the true logistics of life.[3]

Yet this is perhaps too easy a reversal, arguing against a utopia of velocity that never claimed to be one in favor of his own utopian thought that here pretends it isn't—never mind the fact that elevators are the likeliest site for humans to experience the mechanical modification of gravitational forces.

Even so, this lecture speaks to Soleri's rhetorical sensibility, the grounding of megastructural thinking on government-scaled technoscience, and the centrality of such narrative rhetoric in promoting urban ideas—and to the fact that in the early 1970s, it seemed that arcology might legitimately contain the possibility of its realization. The year before, Soleri had been honored with a publicity-garnering exhibition of his arcologies at the Corcoran Gallery in Washington, DC, highlighted by the dramatic model of his "3-D Jersey" project. The US Department of Housing and Urban Development, under an urban renewal demonstration contract, granted $10,000 to the printing of a catalog. Then-Secretary of Housing and Urban Affairs George Romney described this as part of "a continuing HUD endeavor to support institutions in their efforts to bring outstanding new design work to the attention of

**Construction of the "Crafts III" building at Arcosanti, 1970s. Courtesy of the Cosanti Foundation.**

3 Soleri, "The Future: The Flight from Flatness," 31–32.

Soleri being interviewed by CBS News in front of the model for his 3-D Jersey project, 1970. Photograph by Ivan Pintar. Courtesy of the Cosanti Foundation.

A building at Cosanti now being used as a shop and visitors center. Photograph by the author.

the public."[4] Federal investment in imaginative proposals for housing—that alone sounds like science fiction to the contemporary observer.

Soleri took 3-D Jersey seriously as a possible urban form, and unlike many of his later arcologies, he saw it as a project of corporate and political organization as much as a project of speculative design. It is usually said that the project was pursued in partnership with the Ford Motor Company and Rutgers University (though archival evidence suggests that the collaboration never came to fruition), and Soleri maintained a long list of corporate partners—US Steel, IBM, Boeing, GE, Bell, American Airlines, Alcoa—who might be persuaded to join the team.[5] Even a company as staid as Prudential was using Soleri's arcologies in an advertising campaign after underwriting the Corcoran show. His tireless pursuit and promotion of 3-D Jersey, in particular, signaled a faith that the ideas might move beyond his inimitable drawings, painstakingly drafted by his various apprentices and sometimes rendered in charcoal on scrolls the length of a room.

The moment passed, for 3-D Jersey and the megastructure more generally, as the political impossibility (indeed undesirability) of concentrating administrative authority so centrally became clear, and as the federal government stepped back from its Great Society ambitions. But the turn from the 1960s to the 1970s nonetheless cemented Soleri's public stature and marks something of a before-and-after hinge in his work. After a sojourn at Frank Lloyd Wright's Taliesin West in the late 1940s and a move back to his native Italy in 1950 (with his wife, Colly, the daughter of an early client), Soleri returned to Scottsdale, Arizona in 1956. He spent the next dozen years establishing a bell-casting operation (which continues today) at Cosanti, a rambling compound of his workshop's experiments in earthcast concrete, domed apses, and the vaguely biomorphic formal gestures he became known for. One version of Soleri is prominent in these environs—the idiosyncratic desert craftsman, fabricating charismatically bespoke objects out of mineral materials.

---

4 This news necessitated an exclamation point in the American Association of Museums' newsletter. "Department of HUD Funds Corcoran Gallery of Art!" *AAM Bulletin* (March 1, 1970).

5 "3-D Jersey Original Written Materials" binder, Soleri Archives, Arcosanti.

But it was the years around 1970, as he entered his fifties, that the second Soleri emerged into public view: Soleri the urbanist and inventor of cities. *Arcology: The City in the Image of Man*, which continues to define his reputation, was published in 1969; the Corcoran show followed fast on its heels. This was the context in which Soleri and his cadre embarked on the construction of Arcosanti in July 1970. Though the premise of arcology is that it could ground both versions of Soleri in a single manner of thought, these two poles of his work are in other ways strikingly irreconcilable, with craft and scale pulling against each other as a central motivation. The former could be represented by the delicate apses and skeletal forms of Cosanti, the latter by the infrastructural 3-D Jersey—though certain formal signatures trace through both, the differences in their means of realization far outweigh any continuities. Arcosanti can be seen as his attempt to subvert that binary of craft and scale, not through form but through social configuration, with each resident becoming something like a citizen-builder. Despite its pretense to being a laboratory or a new urban prototype, the potential replicability of Arcosanti was always strictly rhetorical, a story told in the name of drawing a community of citizen-builders to the site. What commenced in 1970, then, was not only the construction of a building but the construction of a narrative.

To say that Soleri's reception has generally followed a template would be an understatement. His critics wax poetic on the structures at Cosanti, reading them as they imagine anthropologists (if not paleontologists) might, as evidence of a different kind of civilization. They faithfully describe Soleri's ideas with just enough metaphor to render them folksy; the word "visionary" is never not used. They quote Genesis.[6] They rarely fail to describe the man himself, a wiry, slight, enigmatic object of fascination

**6** Soleri invites this theological bent with his own texts. See Dana F. White, "The Apocalyptic Vision of Paolo Soleri," *Technology and Culture*, vol. 12, no. 1 (January 1971): 75–88; Edward Higbee, "Soleri, 'Plumber with the Mind of a St. Augustine,'" *AIA Journal* (1972); and especially James Edward Carlos, "The Community of Souls: Arcology/Theology," *St. Luke's Journal of Theology* (1971): 45–53.

with a childlike streak. The final paragraph of many reviews of the Corcoran exhibition read like a rallying cry, describing how Soleri was fixing to put the plans into action at Arcosanti (an early model of which was exhibited at the Corcoran). Even the resolute Ada Louise Huxtable found herself enamored of this "labor of love," rendering its optimism as a kind of romantic stubbornness:

> He has bought 800 acres of land 70 miles from Phoenix for Arcosanti, and must meet a partial payment of about $50,000 by June. He has no money... To build Arcosanti, he now has six shovels, some rakes, a cement mixer, some stout-hearted graduate students, and a firm intellectual conviction. Arcosanti is to be a "self-testing" urban environment. If it is a dream, it is the very best kind.[7]

There are dissenters, to be sure—a reviewer in *Landscape Architecture* identifies Soleri's work as a kind of "containerized megamadness" (though the same review arrives at the rather grim pronouncement that what Soleri has made clear is that the only alternative is "a limitation on population").[8] But the Soleri discourse industry revolves largely around believers and interested observers. Even in the later years, when "what was left had curdled into the sluggish but pleasant pace of a nonprofit foundation," as a 1998 workshopper named James McGirk put it, the man himself carried an aura that belied the ways in which The Idea had begun to feel like an anachronism. "He was spry and leathery, and against the sun-baked concrete swoops and apses and arches and circular doors, he looked like a character from a J. G. Ballard short story, the caretaker of a long-dead monument," McGirk recalls. "He had us all squat with him on a mat in the planning room. We could ask him questions."[9]

And there were surely questions, if the workshoppers could figure out where to begin. Soleri's sketchbooks contain familiarly

---

**7** Ada Louise Huxtable, "Prophet in the Desert," *New York Times*, March 15, 1970.

**8** Neil H. Porterfield, "Arcology: A Simplistic View of Man Apart from Nature," *Landscape Architecture* (April 1971).

**9** James McGirk, "Remembering Life in Arcosanti, Paolo Soleri's Futuristic Desert Utopia," *Wired*, April 11, 2013, https://www.wired.com/2013/04/arcosanti-paolo-soleri.

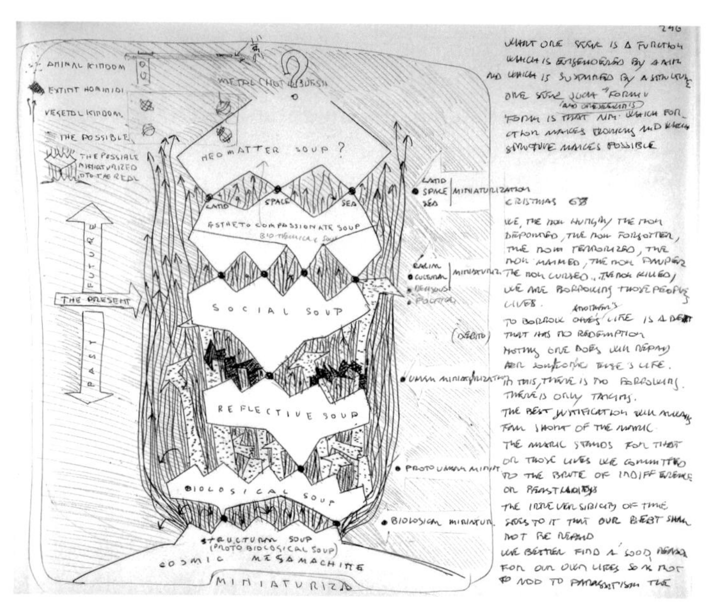


**A diagram of Soleri's philosophy of miniaturization, from his sketchbooks (sketchbook 6, page 246). Courtesy of the Cosanti Foundation.**

architectural kinds of exploration—site plans and sectional sketches adorned with the olive trees and Mediterranean cypresses that would eventually take root on the mesas of Arcosanti—but his thought and public pronouncements were also elaborated in these sketchbooks in a cryptic but all-encompassing language. A characteristic diagram of his idea of "miniaturization," a term that guided much of his work despite its megalomaniacal scale, includes indications for past, present, and future; the animal kingdom, the vegetal kingdom, extinct hominids, and "the possible"; and a series of "soups," whether biological, social, "esthete-compassionate," or "neo-matter"—with all of this grounded on the "cosmic mesamachine." A less enigmatic version of the same can be found in the "Announcement of the Inception of Arcosanti," a text prepared in spring 1970 as Soleri began to formulate the workshop structure that would bring students to the site. "The Arcosanti thesis is simple," his text reads in distinctive Copperplate Gothic, the preferred font when Soleri's wife, Colly, would type documents on their IBM Selectric. "Arcosanti proposes an urban doctrine and will make itself the guinea pig for it. Arcosanti will be a town laboratory engaging its makers and users in the verification of the subject of their learning."[10] Here, Soleri's philosophy is distilled into a somewhat more plainspoken variant, guided by four principles:

> 1. The city is a bio-mental organism contained in a mineral structure.
> 2. The city is an organism of a thousand minds.
> 3. The city is an organism in a constant process of growing complexity. Nature shows that for all organisms or societies of organisms with any increment of complexity, there is a corresponding spatial and durational contraction of his functions! MINIATURIZATION AND COMPLEXITY ARE MUTUAL FUNCTIONS…
> 4. The enormous tide of complexity working itself into the social organism of today makes a correspondingly enormous contraction of its urban container mandatory…
>
> The miniaturization rule, a direct injunction of the logistics of matter-energy, must be respected universally.[11]

That this organicism insisted on being at once biological, organizational, architectural, and metaphysical explains something of the breadth and abstraction at play in his writings and teachings.

It is not surprising, then, that the narration (or perhaps narrativization) of Arcosanti relied on public testimonials from others—and that one of Soleri's greatest skills was attracting chroniclers. The most prolific of these in the early years was the writer Richard Register, whose name shows up with startling frequency in Soleri's clippings file, placing stories with the *Los Angeles Times*, the *Los Angeles Free Press*, the *Arizona Republic* ("What Are We, Men or Frogs?"), the *Philadelphia Inquirer*, and the *Detroit News* in 1970 alone. But Register's most substantive, if less widely distributed, contribution was a 1975 manuscript titled

**10** "Announcement of the Inception of Arcosanti, Spring, 1970," binder of the same name, Soleri Archives, Arcosanti.

**11** "Announcement of the Inception of Arcosanti, Spring, 1970," 3. The original reads "contraction if its urban container" and has been corrected here for clarity.

"A New Beginning" (later published in 1978), which proselytizes for arcology while providing a detailed—and immensely partisan—account of Arcosanti's earliest days. Register is interested in the ideas, translated into a more straightforwardly environmentalist stance on urbanism. He is interested in the man, comparing Soleri to a mythical "first architect...reading between the structural elements like reading between the lines of a story," in which "all the ghosts of the future could be seen waiting." He is interested in the "first citizens" of the place, especially those "paying $350 for the privilege of working six hot, dusty, sweaty weeks in the sun at manual labor."[12] But he is most interested in providing a kind of scenography to the origins of Arcosanti, rendering it in literary and picturesque form.

In a chapter simply titled "July 23, 1970," Register records the first day of work on the site. "Towering white thunderheads were already rolling up in the clear dark blue sky and beginning to rake their rumblings back and forth across the desert mesas, hills, and mountains; an appropriately dramatic beginning, I thought, as we walked down into the valley to start setting up the permanent camp for the construction." During the afternoon working shift, after Paolo's 1PM siesta, another cloud rolls in.

> The sun angled in underneath and kept us bathed in sunlight while the cloud drenched us with a ten-minute cloud burst. We collected the fresh rain water trapped in the sagging plastic roof which we had just erected, and drank a toast to the new city... It was the first day of a new age; I was absolutely certain of it. The rainbow assured me of it ...The solution seemed so close you could reach out and touch it.[13]

This messianism is both a product of, and producer of, the narrative that surrounds Soleri and Arcosanti. Register was well aware of this, his text crafted to serve precisely that purpose. "A word about myths," he writes in an early chapter that reimagines the Adam and Eve story. "The concept of myth used here does not mean fantasy, fairy tale, or false belief, but rather a rich fabric

---

**12** Register, "A New Beginning," 1975, manuscript, Soleri Archives, Arcosanti, 42, 45.
**13** Register, "A New Beginning," 43–44, 47.

of legend and reality that gives direction or coherence to a whole people...A myth vacuum is a dangerous place, culturally somewhat analogous to depression in the individual; a time of confused directions and ill-considered and frequently self-destructive decisions."[14] The supposedly unifying power of myth—an idea that refuses to consider the question of whose myth counts, or how one negotiates between myths, or whether the concentration of myths with singular people and places is advisable—is at the center of Soleri's endeavor, with the participation of the Registers of the world. Arcosanti was always meant to be precisely a myth in that sense, a myth that brings Soleri and Arcosanti into a kind of seamlessness, a tale of conversion and conviction with a wandering band of apostles to spread the good word.

◆

From the northern edge of Phoenix, after passing suburban enclaves with placeless names like Happy Valley and Anthem, it's about forty-five miles up I-17 to Cordes Lake. I pulled enthusiastically onto the interstate shoulder to photograph my first saguaro cactus in the wild, a decent specimen, before realizing the obvious a mile or two later—this is a landscape littered with them. The road climbs from something like 1,000 feet in Phoenix to 3,700, from a landscape of alluvial deposits through sedimentary rock formations up into the volcanic basalt that undergirds the mesas around Agua Fría. A right and a quick left and the rental car rumbles over an old cattle guard onto the premises of Arcosanti. In 1971, when some of the workshop participants were undertaking self-guided projects, a workshopper named Sheri spent time trying to build a replacement for this basic bit of infrastructure. "The cattle-guard did not work, but it was a beautiful job—even the cattle liked it," reports that summer's journal.[15] The driveway passes a sign reading "Arcosanti: An Urban Laboratory," but as the radio cut off with the engine and I stepped out into the sunbaked gravel parking lot, it was not urbanity but a kind of stillness that greeted me.

**14** Register, "A New Beginning," 6–7.
**15** "Jerry," Arcosanti journal, April 22, 1971.

Silence, it turns out, is one of the devices that shapes Arcosanti. As one of the staff members took me around the site, we passed the apse where bell casting takes place just as the pouring of molten bronze began in front of a resident-guided tour group (and the pouring always seems timed with the arrival of a tour group). A hush falls over the assembled audience as the bronze is poured with monastic rhythm and rigor. "A bit of ceremony," my guide whispers to me. The underpopulation of the complex lends itself to solitary moments, certainly, in contrast to the well-honed performances of craft and construction that occupy those same places at other times. But there is also a lingering sense that the current occupants are following a script that was penned some time ago. Arcosanti—like all architecture, though more polemically than most—is as much a carefully designed schema of participation and authority as it is a place. This social character is a part of the architectural vision, with the buildings serving as a reminder of their guiding authority.

Someone once asked Soleri, during one of the public salons he hosted from time to time, who he'd voted for in the Bush versus Gore presidential election of 2000. As the story goes, he replied that he didn't vote. Why? He didn't like it that his vote counted the same as everyone else's. During my own visit, which took place during the first year of the Trump presidency, I spent a meal talking to a resident who had arrived there after becoming disillusioned with American politics. He described how he and many others came to this outpost in the Arizona desert because of their political commitments, and only later found that Arcosanti itself, by dint of its removal from city, by dint of a creeping fixity across the decades, by dint of an insistence on a particular kind of order, ends up resisting the practice of politics, urban or otherwise. And indeed, as my interlocutor memorably put it, Arcosanti is a place that attracts those who have the luxury to be surprised at the state of American politics—a luxury not afforded to many of those who live in the urban settings that Soleri sought to reinvent. And yet, he continued, optimism is a muscle, and he came to Arcosanti with the intention of exercising it. This is one of the qualities of the place, or rather the people who populate the place. The idea of optimism as a kind of practice permeated those who came to join the experiment in the earliest years, and it seems to persist.

Still on East Coast time, I woke up well before sunrise each morning and walked up the mesa just south of the main Arcosanti complex. One morning I disturbed a family of deer who'd bedded down in the grasses; on another I was greeted outside my door by a roadrunner. These things are not a part of my daily life, nor is the cool air of a predawn desert. The first human sound other than my own would be the footsteps of whoever was on breakfast duty, walking up from the camp in the valley—a collection of concrete cubes that date from the earliest working seasons on the site—to get a head start on preparations, soon to be followed by the distant sounds of early stirring. Then the sky warms and the sun skims across the mesas, casting Arcosanti's vaults in a light that makes them appear out of time and place, a quasi-Roman ruin in an unfamiliar geography, an effect that explains the perhaps apocryphal tale that Arcosanti was among George Lucas's inspirations for the *Star Wars* planet of Tatooine. It's possible to know that you're inhabiting clichés and still feel them—as dispassionate as I intended my research trip to be, I found myself recapitulating something like Richard Register's encounter, down to the desire to put animals and weather into what could surely be a less self-involved account of things. This simultaneity of presence and distance—or perhaps believing and doubting—is part of what makes Arcosanti such a complicated place to come to terms with, a complication that was there from the beginning and remains still. Perhaps a visit to Arcosanti asks you to indulge in the "what if" of the project—not Soleri's "what if" so much as a kind of possibilitarianism (to borrow a term from Robert Musil) that's always up for redefinition.

To what does one owe this experience? Neither the genius of an architect nor a *genius loci*, though those are both answers that have tempted many critics and visitors across the past almost-fifty years. Nor does Arcosanti qualify as a kind of land art, framing the natural world anew. Nor does the communality of daily life draw most visitors into imagining their own participation. What snaps you to attention, I think, is something closer to tautology—you've arrived at a place that's been designated as an "urban laboratory," despite the utter absence of anything resembling either urbanity or rigorously scientific research, and your mindset is primed to be attuned to the difference between this and your own life, to view yourself as a potential test subject within this supposed laboratory. This is the narrative of Arcosanti doing its work.

◆

In November 2017, an essay written by Soleri's daughter, Daniela, began to circulate on the internet. It is a devastating but clear-eyed account of her molestation, beginning in her adolescence, and eventual attempted rape at the hands of her father, Paolo—a "fierce narcissist" with a petulant sense of sexual entitlement in private that was masked by the gentle charisma and restless intelligence that he displayed to the world. This is not just a family affair. We have seen the way the esteem around artists and others in the public eye secures the silence of those who have been violated, and as authors of architectural discourse—a discourse that has collaboratively made that character "Soleri," the celebrated and idiosyncratic "prophet in the desert"—we have our own inventory to undertake. For the victim, "sorting it out turns into a threesome," she writes. "You, him—and it is usually a 'him'—and his work."[16]

The Cosanti Foundation has since issued a statement in support of Daniela Soleri—a statement that came six years after her resignation from the foundation's board because of this abuse—maintaining that it "separates the man Paolo Soleri from the ideas, regrets the failings of the man, and honors the ideas."[17] Daniela Soleri writes poignantly of her own fantasy of separation, that the monster and his artistry, which she was still able to find beautiful, might be somehow sequestered from each other:

> I used to dream the same thing over and over. I am a child at home, and there in our living room is my father, Paolo Soleri, in a large cage, fuming. We, my mother and sister and I, quietly hand paper, pencils, crayons, and charcoal to him through the bars, or we hand in clay, or Styrofoam and a woodburning tool, or large flat trays of moist, densely packed silt with knives to carve it, powders and washes to color it.

---

**16** Daniela Soleri, "Sexual Abuse: It's You, Him, and His Work," *Medium*, November 13, 2017, https://medium.com/@soleri/sexual-abuse-its-you-him-and-his-work-88ecb8e99648.

**17** John Walsh, "#METOO," March 8, 2018, http://arcosanti.org/metoo.

He draws, forms, carves, shoving the beautiful results back out angrily, yelling his fury.

But these kinds of cages cannot be built, within a family or within an economy of architectural esteem that has long relied on evading hard truths. "It was a clumsily literal dream...a child's solution to the problem posed by a man who I, and everyone around me, saw as the center of the universe."[18]

That there should be dark secrets lurking here, as there are in so many places and histories, is a truth whose exposure should be welcomed. Arcosanti is a part of that darkness, not only because the early phases of its construction occurred during Daniela's adolescence (she seldom appears in the documents of the period, though she was there on the first day of work at the site).[19] "His work" is what garnered him recognition and something of a guru-like status, but architecture is no inert thing. "His work" is also the organizational structure that he assembled to make it all happen—the very possibility of a self-built arcology links architecture-building to institution-building. "For me," Daniela Soleri continues, "this was the grand flaw that plagued Soleri, his work and his organization: his single-minded control enforced by the deference of all to him...Ultimately, for better and for worse, it became a group held together by their commitment to him, or more accurately, what proximity to him did for them."[20]

Creating that commitment was as much a part of the architectural craft as the structures themselves, providing us with yet another illustration of how concentrations of prestige, authority, and vision might conceal a kind of viciousness at their core. Arcosanti looks different in this light—and it's important that we be unflinching in making use of this light, and this building, and the histories that Paolo Soleri's architecture housed, lest we run the risk of meeting one silence with another. Daniela Soleri's story is not one that runs parallel to Arcosanti—it is another thread in narrative knot that we pick at when chronicling and re-chronicling the architect and his architecture. Narratives change, and in a

---

**18** Daniela Soleri, "Sexual Abuse: It's You, Him, and His Work."

**19** Richard Register, "A New Beginning," 1975, typed manuscript, Soleri Archives, Arcosanti.

**20** Daniela Soleri, "Sexual Abuse: It's You, Him, and His Work."

different sense than its original intentions, Arcosanti now has a new role to play in how we imagine the betterment of our world. "Truth is hopeful," Daniela Soleri writes, "and inevitable."

We need not have waited for Daniela Soleri. While those acts of violence were known only to a few until recently, there has always been much to say, and much left unsaid, about the authoritarian strain that undergirded Soleri's socioarchitectural experiments. In a lengthy profile in the *New York Times Magazine*, published on July 26, 1970—three days after the construction of Arcosanti began—the author relays a story about "a couple of emissaries from Timothy Leary's tribe of acidheads" who arrived at Cosanti "to enlist Soleri... as their architectural guru." Soleri's wife, Colly, proudly describes how Paolo ran them off. "This is no commune," she says. "It's a dictatorship in which Paolo's word is law." The author goes on to tell of "a few militant types" who, in the summer of 1969, "tried protesting the 'undemocratic' working arrangements."[21] That stories like these rarely altered anyone's opinions of Soleri's genius, despite being aired in as public a forum as the *New York Times*, points to how unsurprising it is that architectural practice has long been a province of small-scale despots (in that regard, Soleri learned from one of the finest in Frank Lloyd Wright) and that the discipline has long relied on structural hierarchies of all kinds for keeping its labor in line.

But the workshoppers kept coming, and Arcosanti is theirs too. Among the items in Arcosanti's archives is a journal that was kept on site over two months in the summer of 1971, a remarkable document that offers another vision of what the place actually is—a community of people, some there for the long haul but most just passing through, some paid for their work but most paying $350 for the privilege of donating their efforts to the cause, each struggling with their own vision of the experiment. In this journal, seven voices emerge to tell something of a counter-history of the early making of Arcosanti, a counter-history of the nascent city in which Soleri appears

---

**21** S. D. Kohn, "Paolo Soleri Thinks Very Big," *New York Times Magazine*, July 26, 1970.

as both presence and absence.

On March 15, 1971, a new cadre arrives for the second year of the Arcosanti workshop. Their first two days are spent in an orientation at Cosanti—including a screening of KAET-TV's documentary film *Paolo Soleri*, in the so-called Pumpkin Apse, and a meeting with a zoologist and a geologist—before decamping to the site around noon on that Wednesday. The journal of this working season begins matter-of-factly enough, with a simple summary of the projects undertaken, written in architect's script with a purple felt pen: "March 17. First day at site; pits on mesa for soil engineer; trench for water line; general clean up in batch plant; set cube forms; gravel from river, work on crane shed on mesa." The first setbacks come quickly. "March 18...Cubes moved (2 walls). Footing failed; general clean up. Water line. Paolo up with film people. Will do cubes sanely." That Soleri's presence in the workshop and on the site is often mediated is a leitmotif in the years before the completion of his own house on the mesa. "March 19. Crane maintenance; set up cube forms; gravel up; kitchen repair now OK. But food hurting; possible mutiny but if that's all that's bothering everyone no sweat. Shed on mesa hitting rocks."[22] The weekend comes mercifully, one imagines, with the workshoppers dispersing back to the city or for their own weekend travels. The chronicler from the first few days musters up just three more entries the following week before their hand disappears from the journal's pages.

The journal revives on April 12, four weeks into the first workshop and a week after fresh reinforcements arrive (known in the journal as the "freshman workshop"), this time in the hand of one the participants. "Words are important," begins this first entry penned by one Jerry. "Sometimes concepts can live or die on the choice of words. This is basically a 'construction gang.' But at the same time it is a community. We are building to see an idea become a reality." That community emerges in Jerry's first entry, seven pages long, as a cast of characters—some forty individuals with their own personalities, talents, and shortcomings to be teased about, from Billy's "good hash" to

22 These entries, and the ones that follow, are found in a binder titled "Arcosanti Journal 1970–1971" in the Soleri archives at Arcosanti.

**Construction of Ceramics Apse at Arcosanti, 1970s. Photograph by Ivan Pintar. Courtesy of the Cosanti Foundation.**

Cory's "inept" plumbing, from Anita's tendency to "make other people feel good" to Joe's perpetual lack of letters when the mail comes. The entry makes some attempt at documenting the day's progress, though Jerry is more intent on registering the crew's various enthusiasms and qualms. Despite a number of "questions and doubts," the journal reports that "there are more positive than negative feelings. The reason for this seems, still, to be the work. It's almost an enigma." Jerry ranks the food as the next most important aspect of morale, while "the idea, Soleri's idea, does not rate even second-best in the real-life, nitty-gritty, day-to-day life here at the site." Perhaps this is not just one observer's reluctance to valorize the figurehead of the place, but a largely unheeded warning to the architectural historians of the future—a reminder that the experience of architecture rarely aligns with its intentions and is never single-minded. "Criticism on my writing thus far," Jerry writes, bringing this entry to a close. "Tomorrow, an attempt to be less opinionated."[23]

Jerry is the most prolific, and the most critical, entrant over

**23** "Jerry," Arcosanti journal, April 12, 1971.

the next ten days—believer enough to be there, doubter enough to cast aspersions freely. Soleri's defenders respond in kind. An anonymous contributor writing in a calligraphic serif describes a vision of Arcosanti grounded in humanism, happiness, and above all love, rebutting Jerry by arguing that "we must have an *idea*, a direction in the future to work towards. Soleri has an idea and it is a good form to work towards."[24] The most cramped of the handwriting, eschewing the use of capital letters like a proper modernist, advocates against critique altogether: "i have almost always found that when i criticize or think negatively about someone or something that the fault lies in me, not in those i criticize... keep the faith, this project is just beginning."[25] Others simply articulate their own vision of what the experiment is about. Sheri, who worked on the cattle guard ("I have held a personal grudge against that first barbed-wire gate ever since I arrived"), describes her hope for a city of personal freedom and spiritual warmth. An environmentalist New Yorker reports in a tight cursive on an "inconsistency in terms of the idea" of Arcosanti "and the reality of non-polluting."[26] Dick weighs in with a terse, two-sentence request for a mailbox at the Junction. "Has to happen eventually."[27]

A workshopper named Emile is the most reflective contributor to the journal, dwelling particularly on the implications of life in the high desert. "April 19th. It's interesting how weather affects attitudes. Today was wet and cold and I think everyone felt it. It makes you wonder just how important geo-politics could be." Weather reports make constant appearances in utopian thought, and Emile's climatic view of the project observes the way that taking care of the self—food, comfort, trust—is at the core of redefining life. Immersion in Arcosanti means being oriented away from the world outside, with all of its complications and politics, and toward the knowable hierarchies and the climatic cycles of the immediate present. In that regard, Emile has perhaps already answered the question that follows on this thought: "Why does Soleri choose to live in the desert and to build his prototype there?" ("if anyone knew that they would be the genius and

---

**24** Anonymous, Arcosanti journal, April 13, 1971.
**25** Anonymous, Arcosanti journal, April 21, 1971.
**26** Anonymous, Arcosanti journal, April 18, 1971.
**27** "Dick," Arcosanti journal, April 20, 1971.

not soleri," interjects the decapitalized hand. "ask paolo." Emile responds in the margins: "OK—I will. But I doubt if I'll get an answer.")

But Emile's question is a larger one, cutting to the core of what it means to design a city:

> One of the most frustrating things I have found since coming here is the inability to communicate with Paolo and many of the people here. I think a lot of this is because Soleri and most of those here are too close and too intense about the concept of arcologies to discuss objectively it and *other* alternative futures. Perhaps this intensity is needed to work here—perhaps to question the basic premises now would be like arguing with the captain a hundred miles out to sea, but for myself, I'd hate to see any system become the *only* alternative for the future. Diversity and variety and even the unknown and disorderly are positive aspects in a society. It is not disorder that causes the decay of a society—it is the need for complete order and tranquility that drives people away from the "liveliness" of life into the suburbs and slowly insane.[28]

The metaphor—that these citizen-builders might have embarked on an expedition of sorts that demands nothing short of wholehearted and unthinking commitment—is a vividly romantic one, echoed by any number of Solerians, but one that suppresses the dream of other alternatives, other beginnings. Emile resists that act of aggrandizement, offering perhaps the clearest testimony on record that some at Arcosanti imagined not only a different outcome but a different structure for arriving there.

Soleri is something of an abstraction in these entries, an imagined body around which a set of varying figments of The Idea orbit, though he occasionally enters more directly as the object of Jerry's wry scorn ("the first building is naturally going to be Soleri's home").[29] Only one visit to the site is recorded while the workshoppers are in control of the journal, with Soleri

**28** "Emile," Arcosanti journal, April 19, 1971.
**29** "Jerry," Arcosanti journal, April 12, 1971.

**Paolo Soleri during construction of the individual siltcast panels for the South Vault, Arcosanti, 1970s. Photograph by Ivan Pintar. Courtesy of the Cosanti Foundation.**

questions of him. Essentially he is putting himself at our disposal. He becomes frustrated and impatient when this opportunity is not taken advantage of — that is, when no one has any questions. Tom said something interesting in around this time. He said, essentially, that he feels little, if any, rapport with Paolo. Paolo explained that the briefness of the workshop plus his not very sociable; introverted personality, ~~all make~~ plus his work, all make the possibility of setting up a strong rapport hard to achieve.

I asked my same foolish question, Paolo became his childish, Italian self, called me a few things and succeeded in showing that close-minded side of himself that isn't consistent with a visionary social prophet. This had the expected effect. Soleri has, momentarily atleast, lost some of his glossy finish in the eyes of some of the community, particular among the freshman workshop.

Jerry

**Journal entry by "Jerry," a participant in the Arcosanti workshop, April 12, 1971. Courtesy of the Cosanti Foundation.**

employing his usual tactic of making himself available for dialogue. There were few questions—and here, in Jerry's account, Soleri "becomes frustrated and impatient when this opportunity is not taken advantage of." At one point Tom says that he feels "little, if any, rapport with Paolo." Soleri responded that "the brief of the workshop plus his not very sociable, introverted personality, plus his work, all make the possibility of setting up a strong rapport hard to achieve."[30] This episode draws out the ways in which one might imagine Arcosanti to be displaced from its supposedly authorial hand.

Jerry, however, wasn't ready to take Soleri's excuses as the last word. "I asked my same foolish question," the journal records. "Paolo became his childish, Italian self, called me a few things and succeeded in showing that close-minded side of himself that isn't consistent with a visionary social prophet... Soleri has, momentarily at least, lost some of his glossy finish in the eyes of some of the community, particularly among the freshman workshop."[31] But perhaps that is the greatest myth of Arcosanti, or of any of architecture's many utopias founded by a charismatic leader—that "visionary social prophet" is a worthy aspiration for a designer of buildings and cities, that the subject of a cultivated hero-worship might not be narrow, petulant, quick to anger. This journal shows vividly that while Arcosanti's leader was maintaining something of a studied distance, there was a great deal of genuine debate worth engaging in, if Soleri had cared to—debate that is naïve or experienced, pragmatic or churlish, but always exercising the muscle of optimism. This is also Arcosanti.

In restaging the myths around Soleri the ascetic genius, the "plumber with the mind of a St. Augustine," time and time again, the architectural discipline's writing of its own history has played its part in affirming The Man and The Work, at the expense of seeing it as something unrulier, something that should be reminding us that buildings are rarely actually about The Man or The Work.[32]

---

**30** "Jerry," Arcosanti journal, April 13, 1971.
**31** "Jerry," Arcosanti journal, April 13, 1971.
**32** The phrase is borrowed from Edward Higbee, see footnote 6.

Architecture is an object of the imagination, not only in its creation but in its reception, its occupation, its representations, and its distant cultural echoes—and as we stare down the seemingly intractable tangle of crises that will define the twenty-first century, imagination is much-needed. We also need to understand the architecture of a place like Arcosanti as being evidently more than the material stuff up on the mesa. It is in the journals, the stories, the endless students and workshoppers and passing visitors who've encountered it in their way (as elusive and innumerable as that community might seem), and it is in the violence of Soleri as well. If Arcosanti is indeed the site of a kind of optimism about living ecologically and living together, the very power of that optimism is grounded in being aware of the problems around it. Only by confronting its dark histories and our unsustainable present can we begin to imagine other futures.

To understand Arcosanti this way means questioning the categories to which we assign Soleri's architectural narrations. The subject of Soleri's fiction was not the city of the future, as much as his drawings are interpreted as something of a post-urban fever dream. Rather, the subject of Soleri's fiction was the city as it existed, the admittedly imperfect one he militated against. The ecocidal, carbon-fired suburbias of the "horizontalists" were an easy and righteous enough target, but it was Soleri's willful if unstated narrative around the urban centers of the United States in the 1960s—a vision of unredeemable crime, pollution, bureaucracy, and disorderly complexity—that anchored The Idea of Arcosanti.[33] It was the insight of the workshopper Emile that Soleri's fiction of the city-as-it-existed was wrong, that diversity, variety, and even disorderliness are what give the city its life, that arcology's insistence on an order that refuses to recognize dissent was its defining limitation. What Soleri wrote out of his vision of the city was the same thing he wrote out of his own organization—politics, difference, those things that make cities difficult by design.

---

**33** This vision of the city was rarely articulated outright, though a 1983 article in the *American Bar Association Journal* described an aspiration toward a "crime-free society" by eliminating the "negative stresses" of a "mass urban area." Vicki Quade, "Crime-Free: Arcosanti Awakens in Desert," *American Bar Association Journal*, vol. 69, no. 4 (April 1983): 434.

And yet there was nevertheless a kind of politics being exercised at Arcosanti, though not always the politics prescribed by arcological urbanism or its figurehead—rather, it was the daily politics of living together and contesting ideas. Emile's attunement to the elements, to the pursuit and sheltering of collective life within a climatic frame, is a vital figuration of mutual care that allows us to see a different Arcosanti in absence of its designer, and to imagine others. The journal comes to a close on April 22, 1971—the second Earth Day—with a last entry from Jerry, who, it seems, was turning thirty. Work ended early for a party, to celebrate that and the going-away of many now-seasoned workshoppers after six hard weeks. "The party included ice cream, slides, bonfire."[34]

**34** "Jerry," Arcosanti journal, April 22, 1971.

Arcosanti at sunrise, August 2017. Photograph by the author.

# A Year Without a Winter *Fiction*

# A Year Without a Winter

*Fiction*

*Edited by*

*Brenda Cooper, Joey Eschrich, and Cynthia Selin*

TWO HUNDRED YEARS AGO, *a group of English writers bound for a summer holiday on Lake Geneva confronted unseasonably cold and gloomy weather. Confined to indoor entertainments, Lord Byron amused his guests at the Villa Diodati by daring each to write a ghost story. The stories and poems born there—Mary Shelley's* Frankenstein: or, The Modern Prometheus, *John Polidori's "The Vampyre," Percy Shelley's "Mont Blanc: Lines Written in the Vale of Chamouni," and Byron's "Darkness"—evince the effects of a disturbed atmosphere upon the literary imagination at the dawn of a global climate crisis whose cause would long remain a mystery. Halfway around the world in Indonesia, the 1815 eruption of Mount Tambora was the largest in recorded history, a geophysical event with profound cultural consequences. On the two-hundredth anniversary of the Year Without a Summer, as the Tambora cooling episode is remembered, a group of science fiction writers and scholars led by Dehlia Hannah and Cynthia Selin decamped to Arcosanti to restage the "dare" amidst the Arizona mesa and ask what climate change might mean for the literary imagination today.*

# Contents

# Editors' Introduction

*Brenda Cooper and Joey Eschrich*

The best visions of the future almost always rely on resonance with the past. *A Year Without a Winter* does just that as it plucks stories from the infamous Year Without a Summer as a backdrop to explore how humanity might operate in a future where the seasons are so destabilized that winter drops away entirely. The people who lived through the Year Without a Summer didn't understand what was happening or how long it might persist. We do. There is a vast body of knowledge about climate change. We have evidence from the natural sciences, social science, and more abstract mathematics, and we know that the specter of a "year without a winter" will haunt us for generations. The calamities of each year might vary, but all our years for the foreseeable future will be shaped by the damage our past and present decisions have inflicted upon the earth.

Despite these environmental certainties, the stories collected here aren't merely predictive snapshots or futurist scenarios. They are works of literature, and their vitality derives from imaginative engagement with the inner struggles, moral choices, and everyday travails of their characters. It was important to us as editors to publish stories that excelled on their literary merits, not just as pieces of climate edutainment, and that avoided the disaster-porn pitfalls that bedevil poorly conceived natural-disaster stories.

The success of any collective writing project depends on the collaborators. Our most important job as fiction editors was to choose the writers who would join us on this adventure. We wanted strength, diversity, and stylistic variety. So we talked through possible contributors for a long time, and when it came time to ask, we got lucky. All of the participating writers who said yes were on our initial list of great choices. They come from different backgrounds: Tobias S. Buckell came to the United States from the Caribbean and has already published a successful pair of science fiction thrillers addressing climate change (*Arctic Rising* and *Hurricane Fever*). Nancy Kress hails from an American family in New York State, and has been writing award-winning science fiction for over twenty years. She has multiple Hugo and Nebula awards, a special interest in genetics, and a sharp eye for character. Nnedi Okorafor is the first Nigerian-American author to win a Hugo Award. She writes science fiction and fantasy for adults and young adults. Her novel *Who Fears Death* is in development at HBO, with George R. R. Martin serving as an executive producer. Vandana Singh is both an accomplished short story writer and a professor with a background in theoretical particle physics. We both fell in love with her "Entanglement," a story in the 2014 collection *Hieroglyph: Stories and Visions for a Better Future* (which Joey helped create and Brenda wrote a story for). These are special people, and they wrote gripping stories that provide compelling glimpses into possible futures.

The tales were born from diverse writers but also from a common experience over a long weekend when all four writers joined the four who planned this literary-historical reenactment on campus at Arizona State University. There, we spent time with natural and social scientists, historians, and philosophers from Arizona State University, hearing short lectures on climate change and discussing their implications with fellow researchers and writers. Frightening words were uttered, along with a few hopeful ones. What if our oceans stop being an effective carbon sink? What if intense feedback loops develop as the Arctic ice fractures and melts? What if clean alternative sources of energy, especially solar, become economic no-brainers for power companies, politicians, and homeowners? What could Phoenix look like in twenty—or fifty years?

Well-fed on science, the team sojourned into the desert and stayed together at Arcosanti, a compact, eco-friendly concrete city that was the dream of the futurist and architect Paolo Soleri. Our brief retreat was meant to echo the time Mary Shelley and her circle spent on the shores of Lake Geneva during the Year Without a Summer, out of which blossomed *Frankenstein* as well as "The Vampyre," an early influence for Bram Stoker's *Dracula*. Arcosanti was an opportunity for bonding and deep discussion about the quandaries of climate change and the possibilities of a "year without a winter" as a place for stories to unfold.

An entire short story collection could be written about our time at Arcosanti. The architecture itself is science fictional in that "what did people think about the future fifty years ago?" fashion. It is a partly realized dream that will never entirely work but also continues to birth environmentalist ideas and provide sustainable living for an ever-changing population. Workshops and concerts happen there. People live there. So it is both a failed and a continuing dream, and a monument to careful thinking about the future. It is also ponderous and angled and spare, and set down right in the middle of the breathtaking Arizona high desert. It is ensorcelled with beautiful skies and a landscape full of cactus and coyotes. Simply being there created a strange experience outside of our normal worlds.

We spent time together in a room called the Sky Suite, reading poetry and stories to one another and talking about the strange times that we live in. One writer felt ill. Another was so tired they could barely keep their eyes open. Others were fascinated by the things they found in common. We spoke of climate fiction and science fiction and politics, which was an inescapable topic in that fall just before the 2016 election. There was a freak burst of rain, and stargazing later.

We also experienced smaller fauna. Nnedi made friends with a beautiful and huge grasshopper that seemed to be drawn to her, and only to her. She held it in her hand, and it sat and allowed us to admire the curve of its diaphanous wings. A wasp nest was safely ensconced in Vandana's room and had to be swept out. Two of the intrepid fiction editors are mildly arachnophobic. There were enough gray eight-legged creatures the color of the Arcosanti concrete that one of us slept in the car and one of us did not sleep at all and emerged red-eyed and amazed at our survival.

In the morning, we all participated in an activity to identify signals of technological, scientific, social, and cultural change emerging today. We began to limn the years-without-winters that were beginning to take shape in our overstimulated heads. We shared one more meal in Arcosanti's communal kitchen, with its magnificent view, then drove out of the 1970s ecotopia back into the crowded, still-hot-in-October Phoenix metro area. One more dinner together at the tame Desert Botanical Garden offered a delicious respite from our provocative encounter with the wilder landscapes and architecture to which we'd scarcely adjusted. And after that, the writers scattered to their homes with the challenge of creating the stories you are about to read. Both Nnedi and Vandana spent time seeking out scientists and learning even more. Tobias dived deeply into climate data. Everyone did more research as the experiences sank in.

Science is what elevates science fiction above other forms of literature as a way to shock us into paying attention to our shared future. We all live enmeshed in scientific realities, whether we like it or not. Science fiction helps bring the science thrumming in the backgrounds of our lives—from radio waves knitting together the planet to blood pumping through our arteries—into the foreground. The genre apprehends science and technology in their social contexts, as artifacts of and conduits for power relations and human ingenuity and collective decision-making and (sometimes) disastrous hubris. Science fiction is uniquely well-suited to the era of climate chaos: it clarifies that our diagnosis of climate change and our responses to it are all caught up in both science and sociality, our methods and instruments of measurement, in cultural values and priorities, in faith, both in gods and in the veracity of our economic system. Equally important, it invites us into empathy with people whose lives and situations and opportunities and thought processes are utterly alien. The right story can clarify the stakes of climate chaos in terms of human anguish and anxiety and displacement and hope abandoned, hope stubbornly enduring.

These masterful stories testify to the complexities facing humanity in an increasingly chaotic climate. None provide easy answers. All preserve room for hope. Yet the stories are as diverse as their authors. They vary in voice and tone and style, and they are each inviting provocations. Each of them has made us think, and each of them has made us feel. We hope that you love them as much as we do.

# A World to Die For

*Tobias S. Buckell*

Your hunting party of repurposed, cobbled-together and barely repaired pre-Collapse electric vehicles sweeps across the alkaline-rich dust flats of old farmland. The outriders are kicking up rooster tails of dust into the air behind them, their bikes scudding over the dirt and slamming hard into every divot and furrow. Pennants whip about in the air.

You're glad to be on the top of a pickup with suspension, ass in a sling, feet shoved hard against the base plate of the machine gun mounted right up against the back of the cab. You've been an outrider before, trying to balance a shotgun on the handlebars or the bike without wiping out. You didn't like it.

The outriders might get more respect, but there's a reason they wear all those heavy leathers, padding, faded old football helmets, and other chunks of scavenged gear.

"There she is!" Miko leans out from the passenger-side door and bangs on the roof of the cab to get your attention away from the outriders and pointed front. "Get ready."

Up ahead, through the bitter clouds of dirt that seep around the edges of your respirator, is the black line of the old Chicago tollway. You reach forward and yank a latch on the machine gun, pulling one of the large-caliber bullets into the chamber with a satisfying ratchet sound.

The seventy-year-old gun has been lovingly maintained since the Collapse. It has seen action in the Sack of Indianapolis, spat fury down upon the Plains Raiders, and helped in the defense of the Appalachian Line. You look down the sights, ignoring the massive ox horns and assorted animal skulls bolted onto your truck's hood.

Your quarry is ahead. A convoy of trucks pulling hard for places out farther east. Their large, underinflated balloon tires fill the potholes and scars of the old expressway as they trundle on at a dangerous thirty-five miles an hour. It's axle-breaking speed, a sprint across the country in hopes that they can smash any MidWest Alliance blockades without paying import/export duties.

Fuckers. As if they could just roll across a state for free. Now they'll pay a lot more than just a 10 percent transit fee.

"Cheetah cluster: right flank!" Miko screams over the screech of old suspension and the rumble of tires. He is still hanging out of the door, and he points and throws command signs at other drivers. He's ripped his respirator off and left it to dangle around his neck. "Dragon cluster, left. Cougars for the front."

Your cluster of vehicles splits off to swing behind the convoy's dust trail, the world turning into a fog of black dirt and amber highlights, and you fall in on the right like a vulture. Enemy outriders split off from the convoy to harass the impending clusters, but they are outnumbered. Shotguns crack through the air; people scream.

In moments the security around the convoy peels off, uninterested in paying for cargo with any further lives. They've done their paid duty—they can head back in honor to whoever hired them.

"Do it," Miko orders you. He pulls his respirator back on, and now he's all green eyes and blond hair over the edge of the cracked rubber. He's the commander. You've risen far following in his footsteps. Maybe one day you'll run a cluster, give commands to your own outriders. You've been tasting ambition like that, of late.

To survive, you need to find the right people to follow. Miko has created a strong pack. The fees you pull from what little cross-country traffic still trickles over the road keeps you all fed, the respirators fixed, and batteries in stock.

You lean back and pull the trigger.

The old Browning destroys the world with its explosive howl. You rake the tires of the trucks as Ann, the driver, moves the pickup along the convoy in an explosion of acceleration.

All three of the oversize vehicles shudder to a halt. Ann brings the pickup to a sliding stop, dirt and chunks of the old highway rattling up to kick the undersides of the vehicle.

Miko steps out, shotgun casually slung over his shoulder.

Everyone's expecting the drivers to come out of the truck cabs with hands up. But instead there's a loud groan from the trailers. Miko swings the shotgun down into his hands and aims it up at the sound.

The sides of the middle trailer fall open and slam to the ground. Hauz Shäd shock troops in their all-black armor are crouched behind sandbags and a pair of .50-caliber machine guns.

You're all dead.

But instead of getting ripped apart by carrot-size bullets, one of the Shäd shouts through a loudspeaker, "We seek information from you, and only information. Drop your weapons and live. You'll even be allowed to keep them. In fact, give us what we seek and your cluster can take the entire shipment of solar panels in trucks one and three."

Miko drops his shotgun.

You push away from the machine gun, hands in the air, wondering what happens next.

The Shäd jump down from the trailer, greatcoats flaring out behind them. Their deep-black machine guns seem to soak up the amber light. There's a storm brewing up north; you can tell. You will all need to run before it, get down underground before the tornados touch down and begin ravaging everything.

"Remove your ventilators," the nearest Shäd gestures. "Kneel in a row."

Your mouth is dry. This is an execution line and you know it. Bullshit promises aside. People like the merchant riders of Shäd view the continent as a place they should be able to trade across. They view clusters as "raiders" and not the customs agents you know yourselves to be.

But instead of walking behind you, the Shäd spread out in front of the line, moving from member to member. They're holding out photographs in their gloved hands, looking closely at each member of your cluster through mirrored visors.

The nearest Shäd approaches. Those ventilators they wear are not just the usual air purifiers, you realize. They're connected to oxygen bottles on their hips. You've never been close enough to see that before.

"We are looking for someone," the Shäd says through his mask.

*The fuck? All this over an MIA?* People go missing in the middle countries all the time.

"This is her picture. We have been told she may be called Chenra, or Chenray. Have you seen her?"

You glance left, blood going cold, because that's *your name*. Miko stares at the ground and shakes his head. Ann, her long black hair tangled up in ventilator straps, shrugs. Cheetah cluster won't rat you out.

But why the hell are shock troopers from Hauz Shäd hunting you? You've been on a few nondescript battery raids, run the outlander position on some basic convoy stops, but you're not officer class. Just a runner and a gunner from nowhere.

You'd be sick from fear if you even understood any of this.

You're mainly confused.

The toxic air is rasping at the back of your throat. The ever-present dust is making your eyes water.

It has to be some kind of mistake. Whoever they are truly hunting has used your name, or has a similar name, and these very dangerous private security troops ended up crossing the midlands to find you.

"Anyone who gets us this woman can have the solar panels in the trailers here," the man repeats, pushing the photo at you.

"Yeah, let's have a see," you mutter, leaning forward on your knees.

He steps closer and you take a look.

It's you. There's no mistake.

But you know for sure you've never had your hair cut up above the ears like that. Or so flat.

The teeth are all wrong. White. Like someone has painted them. Different positions, too. *Shit, do you have a long-lost twin sister or something like that?* The woman in the picture does look eerily like you.

But it can't be.

A whole shipment of solar panels. If they're telling the truth…

You stand up. "Are you serious about those solar panels?"

A helmeted nod in response.

"What's your business with that woman?" You jerk your chin at the photo.

"A client needs to talk to her."

"That's it?"

"Just a conversation."

You take a deep breath. Cheetah has given you food and lodging. Given you a trade. You were thirteen when they found you trudging across a dune up in the lower peninsula of the Holy Michigan Empire. They treated you well. Better than you'd feared when you saw the motorcycles roaring across the sand toward you.

Even if you get shot, or kidnapped, a solar shipment would help all the clusters here. And if there's one thing you are, it's loyal to the people who show you the way forward to a better life.

"Then I'm Chenra, Cheetah cluster. I claim the reward of the solar panels."

"Che! I'll follow you!" Miko tries to stand, but one of the Shäd casually kicks him back down with a boot to the shoulder.

"You can't come where we're going," the mercenary laughs through her respirator.

***

Hauz Shäd has a strong reputation for following contracts to the letter, so you're not overly worried anyone will get screwed over here. All that crap about them eating the flesh of people they've killed is just rumor. Collapse jitters.

Sure enough, as you're taken up toward the head vehicle, the trailers are being disconnected from the trucks. Several cluster members climb aboard, open the rear doors, and shout in delight. Outriders watch you go by and nod in respect.

You don't see Miko anywhere. He's a ghost when he needs to be. He's not going to be happy about his gunner getting kidnapped.

Clusters don't carve promises into skin and swear blood oaths to protect one another to death as a meaningless gesture. They're planning to watch and follow you, to have your back. So you're going to play along, see what comes of this bizarre attempt to kidnap you.

If you manage to get out of this and back to the clusters, they'll all owe you big. Maybe even get you a promotion. You could end up a driver inside the shielded cockpit of an attack pickup. Maybe even get some scrip for hydroponic fruit from down near Fort Wayne.

The truck at the head of the convoy has an extended cab over the battery frame. The up-armored doors hiss open, and the Shäd on either side of you point inside. Other Shäd are putting new tires on the truck because your Browning has torn them up.

You clamber up and into a sumptuous small office.

The door shuts behind you, and you wait a split second for your eyes to adjust.

Inside it is like something out of an old pre-Collapse magazine. On some small level, when perched on a shitter and leafing through the faded pages, you'd convinced yourself that those photos were fantasies and fakes. But the interior of the back of this truck cab is all clean white leather, glossy polished wood, and black electronics.

There are no spliced wires, jury-rigged equipment, or bolted-on extras. Everything in the interior screams newly manufactured. And the air. It's crisp, cold, and doesn't burn with pre-Collapse irritants. The filters in here have to be brand-new, not salvaged or refurbished. This all has to be from one of the city enclaves, you think. Because no one *makes* stuff anymore. Or maybe things are turning around somewhere on the continent and this truck has been manufactured, not reclaimed.

You remain standing, suddenly hyperaware that bucket seats like the ones around the table back here don't get sat on by dusty road agents like yourself. But the man sitting on the other side, framed by a pair of flickering flat screens showing long lists of data and charts, waves a hand for you to sit.

So you sit.

"I think I know who you are," you say, a little tentatively.

The man, brown hair thinning at the top and showing some gray, his blue eyes slightly faded with time or sun, nods back at you. "We have met."

"You're the gold trader. From the Toledo Bazaar. Armand."

You remember that he'd been overly interested in you when you'd come in to trade gold for solar equipment. At the time you'd written it off as him perving out. You'd stepped back to let Miko handle the weighing of the jewelry, confiscated from various folk attempting to run the toll road without paying.

"I am Armand." The gold trader does still seem interested in you, but he isn't leering. He looks concerned when he leans forward across the table.

"So what's all this about?" you ask.

"Someone is trying to kill you." The truck lurches into motion. You stand up, fear and anger stumbling over themselves as you grab the door handle. It's locked, of course.

"What the fuck are you doing?" You reach for the knife in your boot, something that the Hauz Shäd goons didn't bother to pat you down for. Maybe because they didn't think a woman would have one, or maybe they didn't care. That seemed more likely.

"As I said: someone is trying to kill you," the gold trader says, looking at the knife in your hand but not looking alarmed. "I'm rescuing you before they do."

"Didn't ask to be rescued." You twitch the knife at him. "Take care of myself well enough, thanks."

"But you actually *did* ask me," the man says. He slides the photo, the one that the Shäd had showed everyone on their knees in the line, across the table.

"I don't—" you start to say.

"You did ask me to rescue you. The you that I'm staring at right now."

You stare at the photo. "Nothing you just said makes any sense to me."

"I know." He taps a command out on the nearest screen. Readouts that you don't understand flicker on, replacing the text. Bars representing power levels. Complex math scrolls across other screens. Hieroglyphs you don't understand.

"This will feel weird," he says, and slides his finger up one of the screens.

The world outside the windows inverts. Not upside down, but inside out. It's an impossibility that causes your stomach to lurch and your mind to scream as reality, for the briefest moment, ceases to make sense.

***

A loud explosion rocks the trailer behind the cab. The truck shudders to a stop. Smoke trickles in through the office. This could have been something Miko did, now that the trucks are all far enough away that the clusters are safely running off with the panels.

When you grab the door, it thankfully opens. "Go!" Armand shouts, choking on the smoke.

You stagger down out onto the road, coughing and then retching.

You take a last shuddering breath and straighten up, frowning. The road you puked all over is a seamless expanse of newly poured asphalt. It's faded, the lines are patchy, but there are no major potholes. Or, there were places that had been potholes but were filled in.

The truck has pulled over to the side of the highway, onto a gravelly shoulder. There's smoke pouring out of the trailer, and Shäd are running around, likely trying to stop the fire.

But you hardly pay attention to that. You're staring down the highway, where a car is screaming down toward the group at high speed.

*Shit, shit, shit*, you think. You're all under attack.

But no one pays it any attention.

It whips past, the wind shoving at you, and then with a whine it's gone.

Armand the gold trader is watching you closely. "What else do you notice?" he asks, smiling slightly.

"The air is filtered," you say.

But that doesn't make any sense.

You turn and spot the green. Some kind of short plant covers the dirt, miles and miles of it. You think back to books and magazines you've scavenged in the past.

"Soybean?"

"Genetically modified to handle these levels of carbon and a variety of pollutants, yes," Armand says.

"Did we travel back in time?" you ask, trembling and thinking back to the moment the world outside the cab's windows *inverted*. About half of Cheetah cluster can't read, but you've pored over moldy, yellowing pre-Collapse novels. They're not as valuable as the textbooks, encyclopedias, and practical nonfiction that are near currency, but you've read some freaky shit, and this is the first thing your mind throws up as a possibility.

The wind is cool on your exposed skin. You all wear leathers, but that's for protection against falls and the acid rain if you get caught outside. You're always sweating in them. The cool wind makes you want to strip down and let it play across your skin.

The gold trader smiles. "Well, the air is more breathable here. It's like the air where you were eighty years ago. But it's not time travel. It's a middle RCP world."

"What?"

Armand's faded blue eyes tighten. "RCP: representative concentration pathways. It's a name given to scenarios based on how much greenhouse gas is dumped into the atmosphere. This world right here, it never hit the Collapse. Every world has a different value, depending on how they handled things."

You look around the fields, the smooth highway, and look up at the blue sky. No haze. No impending thunderstorms. A low-RCP world. No, middle, he had said. You can't imagine anything nicer than this. It looks so glossy and early 2000s. "You're like the fucking ghost of What Christmas Could Have Been."

Now you have a faint suspicion of how he could just give away a trailer of solar panels without blinking. This place you're standing in, whatever it is, is a rich paradise compared to the dust bowl of the midlands.

You're half-convinced he drugged you and took you to some sort of promised land, but there are no gaps in your memory. Some of the outriders talk about Edens like this. But they're usually under a dome of some sort. A green, air-filtered paradise in a dust-pocked hellhole of algae farmers and people running around with respirators.

And lots of perimeter security.

Or sometimes they're rumored to be built deep underground, the light all artificial.

Some of the eggheads predict that eventually humanity will just... fade away. The heavy amounts of carbon our forefathers dumped into the air had long since hit greenhouse runaway. The heat being trapped causes more clouds to build up, which in turn causes more heat. Eventually we won't be able to breathe outside. We'll go full Venus.

Maybe after that, it will just be assholes in domes, and everyone outside dead.

But all this is no dome. Not this big. And you are definitely not underground. You're outside.

Another car rushes down the freaking *highway* like it's no big deal. One of the Shäd approaches Armand. "It was one of the capacitors. We have enough spare for the next incursion, but we should really hit up the depot on the other side, or make the trip to a depot here."

"Oxygen?" Armand asks, his face twitching with annoyance.

"Enough to pass through."

"Let's do it. We're on a tight schedule."

You've taken several steps away from the quick meeting. Armand notices and focuses his attention back on you, switching back from a commanding presence to something softer. You instinctively feel defensive. Manipulated.

He smiles at you. "This is going to be a lot to take in, but we don't have much time and there is a great deal at stake. I know you can absorb this all quickly. I've seen you do it before."

Phrases like that, his familiarity with you, are starting to fuck with your head. "Talk," you say, and jut your chin forward a bit. He can play all friendly, but you have an invisible wall up.

Good God, this air is fresh and sweet. You could almost drink it.

"If I know you, and we go way back, you and us, you've gotten your hands on anything you can read. Even in that shitty dust bowl we were just in. That was a universe, right next to this one that we're in now."

It's the sort of thing you talk about to a buddy, lying on the hood of a truck and pushing cannabis through a respirator while staring up at the stars. Imagining that this universe is inside of an atom inside of a cell of a blade of grass inside another universe and on and on. It's great stuff when you're high. The idea that a better, different universe could be an impossible razor's width away if you could *vibrate* over there in just the right way.

But Armand is trying to pitch that it's real. That he's taken you over into an alternate reality. An alternate history. And looking at the rolling fields of farmland, growing crops, you think it has to be true.

You listen to his explanation and ask, "And you cross over with a truck and trailer?"

"Self-contained mobile operations center," he says.

And you're going to ask why it needs to be mobile when gunfire rips through the Shäd milling about the edge of the road. The fearsome mercenaries are taken completely by surprise as ashen-faced Cheetah cluster warriors advance from underneath the trailer and wherever else they'd been hiding.

This isn't the first shipment Cheetah cluster has slipped aboard to fuck up later. Revenge, hijacking, or otherwise.

Shäd fall, shot in the back, and others are tossed from the top of the one trailer behind the massive truck, bodies limp. Armand spins around to take in the ambush, and you take the moment to slip behind him, press a knife against his throat.

"Jesus Christ, Che, this is not a good time," he whines.

Miko jumps down from between the trailer and cab, road dust caking his road leathers. He raises a hand in greeting as you shove Armand forward.

"Where the fuck are we?" he asks. "And what's wrong with the air?"

For a moment, you think about it. Would any of what Armand said make sense to a man like Miko? He only sees what is right in front of him. Profit now means good living now.

You need to find out more about what's going on before you can try managing upward. "Let's get into the trailer."

Miko smiles. "See the salvage?"

There might not be any. Not if there's some universe-crossing engine there.

But you can't imagine that someone would be packing Shäd and crossing worlds with a long trailer hauled behind them if they were just coming to pick you up. There's got to be something else going on. You want to see for yourself.

You prod Armand's throat with the knife enough to draw blood when you reach the rear of the trailer, eyeing the thick doors and the security keypad down at the bottom of the door. "Time to open up."

"This is a huge mistake," Armand says. "We need to be moving along."

His voice cracks slightly, so you believe he's nervous about something. Whether it's about what you and Miko are going to see in a second or something else, you're not sure.

"Open up, or we slit your throat. Then we leave you here and take the truck anyway," Miko says.

Armand swallows nervously. "Che, you know this is a bad idea. Kill me, and you're stuck here."

"Stop calling me Che," you tell him. "That's not my name. It's Chenra."

"You usually like being called Che; it's something of a joke for you," Armand says.

You don't see you and Armand being buddies, no matter what world or alternate reality you were ever both in. "Open up or maybe I take my chances being stuck here. I like the air," you hiss at him.

It's not a lie. Armand can hear that in your voice.

He taps out a code, simple numbers that you memorize, and the doors slowly fall open toward the ground to make a ramp. Miko moves in ahead, pistol in the air and his road leathers creaking slightly as he walks carefully into the dimness of the trailer. You follow, pushing Armand ahead of you.

Your eyes adjust. To the front of the trailer, there's machinery. Pipes and wiring. Shit-tons of wiring. Readouts glowing in the dark.

A pair of Shäd are waiting, weapons aimed right at you. But Armand shakes his head at them, despite the knife you're keeping by his throat.

This is apparently not a place he wants gunfire inside. The outside must have been armored, as he hadn't worried about using the trailer as a lure for an initial attack. But inside...

"Drop them and kick them over," Miko shouts.

They do so and walk out, leaving the three of you alone.

All along the walls leading to the engine are storage lockers.

You recognize the glint of precious metals piled in plastic bins. So does Miko. He's laughing happily, kicking the cage doors open and shaking the bins. "Fucking motherlode," he says. No one in the countryside gives a shit about gold and diamonds. Gold doesn't give you power from the sun, doesn't feed you, doesn't crack out the pollutants from the air. But the traders like Armand in the cities are always looking for it. In exchange they'll give out solar panels or batteries, even food.

Miko stops in front of a cage with old, yellowing paintings. "What's this? These fat-faced people in black clothes."

"Nothing," Armand says from between clenched teeth.

"No." You recognize one of the stacked paintings. "That's a Rembrandt."

Miko takes a closer look at the yellow-and-black hues. "This is bunker shit." He turns back to you and Armand. "You have contact with bunkers?"

The greed makes Miko's face twitch. Dome folk, they know they need perimeter security. Bunker folk tend to get lazy. Think that being underground and hidden makes them safe. Ready for runaway atmosphere with their scrubbers and technology.

But even now they're already having to trade for essentials. Turns out trying to build a balanced ecosystem is a bitch in close quarters.

"I can't give away client locations," Armand mutters.

"Oh, but you will. Eventually," Miko says with a small hint of glee in his voice. The clusters are going to be fat with spoils, and he can taste a new, bountiful future. One that isn't scraps from the old roads.

"You can be much richer, much better off, than that," Armand says to you, almost begs of you. "Think smarter. I told you where we are. He's still thinking of what is important to your world."

Sure, you think. But if this stuff wasn't important to these worlds, why is Armand taking it from yours? The paintings must be valuable. The gold.

"Whoa," Miko says, and Armand's shoulders slump slightly. "What do we have here?"

The last two lockers before the complicated engine and wiring contain people. Two women and a man, shackled together to the wall and wearing gags. They're out cold, small plastic tubes running from their arms up to a device in the wall.

"Slaves," you say bitterly. "You're running slaves."

It's a fate you avoided by joining Cheetah cluster. Why you fight hard. Because things can go grim out in the dust. The country you live in was born on the blood of slaves so many hundred years ago before the golden age. After the Collapse, it had turned, ever so easily you think, right back to it.

"These are famous people in the other timelines," Armand says quickly. "People pay a literal fortune to have a personal servant who is also the president. Only the mega-rich can afford it. These are not slaves; they'll be given more than they can imagine. They'll come around. They always do, when they see the higher RCP worlds. Better to be a servant in paradise than a ruler in a hell like yours."

"What are you talking about? What's an RCP?" Miko shouts back at you. He's standing in front of the bulk of the machinery, gun in hand, not sure what to make of all the wires and readouts. He's fixed a lot of engines in his time, but this is like showing an electric motor in the hubs of his pickup's wheels to a monkey.

"Was that going to be my fate?" you ask. You aren't able to take your eyes off the bodies behind the doors.

Armand's eyes widen. "No, I swear it."

He's holding something back. You can smell it. Armand twists away from you, and you sneer at him. "Is that what it is? Am I someone famous on the other side?"

"It's more complicated than that," he whimpers, seeing the rage take light in your eyes. You can't hold it back; you want to gut him and watch him bleed out.

Miko senses this; he's been on the blunt end of your anger before.

He grins and slides up behind Armand, his dirty leathers a brown-stained contrast to Armand's tailored black suit and shiny shoes. "I'll kill him slow for you," Miko whispers.

He's just messing with Armand. Miko will fight, but only fair. He's a soldier, not a murderer. But he's ridden with you long enough, commanded you long enough, that he can read you. In that brief moment, your long list of ways Miko can annoy you fades: the innuendo, the grabby hands, the little gifts after raids. You know he wants you, over any of the other warren girls who would throw themselves at a commander. He thinks it would be hot to fuck the gunner, and he's been obsessed with that for nearly three months.

"You're a genius," Armand babbles. "You're a genius."

Miko makes a face. You think he's being a little bit of an asshole. But you essentially agree. You're a good gunner, but you're no genius.

But you feel like, thanks to Miko being a bit murdery, that something may have been dislodged from Armand's slimy mouth. "A genius?"

"On one of the other sides, a little further down the line," Armand says. "I don't just transport servants. Or priceless art, cultural artifacts. I also move priceless minds. Think about it: there are minds that are brilliant but trapped by the circumstances of their timelines. They're handicapped from the moment of birth, no matter how great they become in one of the stronger worlds, because somewhere like yours they're fighting just to breathe. They don't have time for their greatest inventions, or to achieve their great works. So we rescue them to bring them over."

"Yo, commander, we got problems," one of the Cheetah cluster outriders shouts from the ramp. You squint. It's Binni. "They got cars with lights on them pulling up, and they got guns."

Armand looks past your shoulder to the outrider. "Don't shoot back!"

"Don't worry trader, we got it. We'll push them back." Binni grins.

"That isn't...they have *resources* here," Armand grits out.

The familiar crack of gunfire fills the air outside. Everyone crouches and moves to the doors. It's there you see the cars, like something from old glossy pages. Smartly painted vehicles, livery matching the official black uniforms the enforcers are wearing.

These are...*police*. Like internal security, but for keeping law and order on a scale that seemed like a fairytale when you read about them.

They're not hardened soldiers but civilian peacekeepers, crouched behind their vehicles as Cheetah keeps them at bay. Binni is right. "Those uniforms can't match Cheetah," you say to Miko as you jump down and get Armand to shut the doors.

"Yeah." Miko agrees, but he's looking more and more pained. All of the things that don't make sense are starting to get to him, you can see. "Who the fuck are they?"

"They're called police," you tell him.

Miko's looking around, more now than before.

Armand all but tries to shove the two of you up toward the cab. "We can't stay. It won't be safe. There are rules to all this."

"Shut up." Miko cracks the side of Armand's head with the back of his metal-studded leather gloves.

Armand staggers, blood dripping down onto his immaculate suit.

"Look," you say to Miko, pointing at the sky.

An aircraft is banking over the farmland toward the road. The blades blur through the air, and a distinct *whump* reaches you.

Miko's attention is fixated on the helicopter, as is yours.

No one has had the spare fuel to launch aircraft in your memory. Batteries don't keep them up for long enough.

Then a second and third helicopter join in.

Farther above, you suddenly realize the long, stringy lines of clouds are aircraft, jets high up in the sky, and you can't help but stare.

Armand slams the door to the cab shut.

"Shit." You and Miko jump onto the side of the cab. Miko puts his gun up to the glass of the cab. "Open the fuck up, or I start shooting."

Armand ignores you both, tapping at his screens. The trailer hums as the massive device buried in its front half kicks on. Miko steps back and fires. The glass doesn't crack, and pieces of the bullet ricochet, clipping him on the shoulder.

You ignore all the shouting from farther back. "Miko, don't get off the truck."

He glances over at you. "What?"

You lean back to shout the same warning at Cheetahs scattered around the trailer and road, along with their Shäd prisoners and dead.

Before the words are even formed, the world turns inside out.

***

You know to expect the sick feeling in the pit of your stomach that causes you to lean away from your grip on the door handle and think about vomiting. But you hold it in. You have to.

But now that you've handled that, a searing stinging in your eyes blears your vision. You didn't expect the crawling pain on your exposed skin and the thick muddy clouds all around you. Wind whips murky clouds around overhead.

The world around you is hell. From the choking heat to swirling, searing moisture.

And you can't inhale the air. You realize that the moment you're exposed. You're holding that last breath of sweet air from the other world, or universe, as Miko stares at you in complete and utter horror. He's choking.

You take a moment to orient yourself, squinting and blinking, then scuttle along the side of the cab. All Armand has to do is wrench the steering wheel to slew the truck about and you'll fly off the side. You dig your shoes into the lip under the doors and reach for any purchase on the side.

Then you swing around behind the extended area of the cab and fall onto the dead Shäd wedged in the hoses between the trailer and cab. You rip the facemask from the body and take a deep breath. Miko's boots strike metal behind you.

You take a last deep huff of bottled air and pass the mask over.

Vomit streams from Miko's chin, and he fumbles for the mask, unable to see anything with reddened eyes. He breathes from the mask as you wait for your turn.

When you get it, the sour smell triggers a second round of stomach clenching.

You pass the mask back and forth as the truck grinds its way over a dirt road for an hour. Sometimes you think you see structures, tall and lurking in the distance. But their edges are shattered, and many of them are slumping over. You might be passing through the edges of a city. In your world, there isn't anything here for another couple hours, but this is a dramatically different reality.

The storm you were dropped into subsides. Maybe it's the buildings blocking the wind, or maybe it blew itself out. You're no longer getting pelted with small pebbles. The ocher clouds overhead still scud by impossibly fast.

Miko's eyes are wide. You're not going to be able to explain science fiction ideas about variant universes overlapping each other while passing a mask back and forth. He looks weak. He inhaled too much of this soupy shit. And the heat is going to drop both of you soon.

This, you think, is a high RCP world. This is what your world will look like at some point. Hothouse runaway, the heat-trapping clouds overhead creating more heat as the whole ecosystem cycled toward something sinister and hellish. Only here they got to it sooner. They burned more fossil fuel and burned it faster, dumped heat into the atmosphere faster. Got here well before you.

And in the other world you were just in, they didn't. Their ancestors somehow restrained themselves.

What did that look like? That restraint?

Could you have done it? Miko? No. He thought only about his next meal, his next fuck, his next raid.

Miko retches hard, his eyes bugging as he cries tears of blood. You give him the mask, even though you're faint yourself. There's dark blood all over his shoulders from the bullet fragments rebounding when he shot the armored glass.

How long will it take for Armand to get to the next jump? He seems to be driving with purpose, trying to find a distinct spot.

Will Miko even make it?

You rummage around the Shäd's body, looking for the bottle of air the mask is connected to. It's dangling by his hip. The gauge is at half.

If the man had a full tank when you jumped, and had enough air to be safe during a transit, then you and Miko are going to suck through it before you're halfway across this side. Right?

Miko shoves the mask at you and tries to lean closer. "I'm sorry," he rasps.

"Save your damn breath," you tell him.

"I shouldn't have let you go with him," Miko struggles to say, trying to push the mask back at you and turning his face away from it when you give it back to him. He takes a shuddering gasp of the hellish, hot air.

"No!" You try to force the mask over his face, and he weakly grabs at your forearm to try to stop you.

The truck and trailer judder to a sudden stop.

You snap your head around, then take a few deep hits from the mask and move around Miko to look down the length of the cab.

There are shadowy forms lurking in the brown mist. Armand hits the lights, and you see a crowd of a hundred or so people in the middle of the dirt road. They are on the other side of barriers made of pitted concrete and rusted rebar.

They're all wearing tattered rubber ponchos, faces obscured by gas masks. Many of them are carrying crude guns; others are holding spears or bottles of fluid with rags hanging from the tops.

Armand must not have any Shäd in the cab to drive, as he spends a long set of moments lurching the trailer and truck into reverse and trying to correct the trailer from sliding to the side. But he stops again, and you see that barriers have been rolled across the road. Sharp rebar is pounded into the ground to hem the truck in.

You recognize the tactic. You've been the one pushing barriers across a road before.

Miko has slumped over and pushed the mask away. You take a few pulls from it and try to get it back on his face. But he's not responding.

Figures surround the truck, surging through the muck floating over the ground. You're trying to pull Miko down when they grab you and pull you forward. The large crowd parts as a figure walks confidently to the front of the truck with a massive rocket launcher over one shoulder.

The person stops and pulls their mask away from their face to shout up at the driver's side window.

"Armand: get the fuck out of the truck, or I'll fire this right at the window."

You stare.

The face. There's a long scar across the cheek to the nose. The hair is shaved down except for a slight tuft near the front. It looks older, more weathered.

But damn it, it's *your* face.

"Hello, sister," the woman holding the rocket launcher says, never taking her eyes off the truck. "I'll talk to you in a few. I'm in the middle of something with that rat-bastard up behind the wheel."

You open your mouth to talk, but one of the poncho-wearing warriors to your side jams a mask over your face. You breathe the clean air in deeply and gratefully, then let them lead you around to the back of the truck.

There's a gun pointed at you, and they force you to sit near Miko. He's not moving, but they've pulled the Shäd mask on over his face so he can breathe. If he's still alive.

Four peel off and pull out what look like jury-rigged explosives that they start taping onto the doors.

You peel the borrowed mask away. "Hey! You don't have to do that."

They pause and stare back at you, bug-eyed in the gas masks and startled at the interruption.

"I know the code," you tell them.

They look at each other, then shrug. Four guns are trained on you as you stand up and slowly walk over. You punch in the combination you memorized when watching Armand, and the door slides down to become a ramp once again.

They swarm in, weapons up, to secure the back.

You hold on to Miko and watch them break open the cages. Miko would be devastated: all that loot's going to end up in the hands of these raiders.

Raiders run by someone with your face.

Another universe. High RCP, you tell yourself. You're in something like shock. Is it the air?

Your self-labeled "sister" arrives, pushing Armand in front of her. He's holding his breath, eyes wide, and blood running down his temple. You're not sure who's having a worse day: you or him.

You get waved up the ramp, along with Armand. Someone helps

you pull Miko up into the trailer, and the door shuts behind you. Air pumps run for a second, and then everyone starts removing their masks.

This other version of you, clearly the leader here, clearly in control of this strange situation, looks you up and down and doesn't seem to come to any sort of conclusion one way or another. "I call myself Che. It's a little bit of a joke, if you read history."

"I'm Chenra. Full Chenra."

"Good for you." She is unimpressed. "Armand explain the smuggling he does?"

"Some of my people rode the trailer, ended up in another world, tried to take the truck," you tell her, not answering her question but trying to explain who you are. And why Armand is your enemy. Can you make friends with yourself if you both hate Armand? "We didn't really understand what was happening, and he took the moment to jump us over here. We didn't have bottled air."

Your doppelgänger looks down at Miko and nods, a riddle solved. "He's in a bad way. How long did he inhale the soup?"

"Long enough. Has something in his shoulder from trying to shoot the window." You nod at the art and valuables being dragged out toward the back of the trailer. "That all yours now? I'm not trying to lay claim to it or get in your way."

"What do you think, Armand? Should I keep all this shit?" Che kicks your captor, who is lying on the floor, spitting bile. He wipes his chin with a dirty suit sleeve and glares at her. "He tell you how special you were?"

You lock eyes with yourself and shiver slightly. "Yeah. He said I was a genius."

"It's not all bullshit." Che pokes Miko with the tip of her boot. "You want to try and get him home to your own universe, to help him, right?"

You nod. No sense in trying to lie to yourself, right?

"You two screwing? I've never seen him before," she says. "Not really our type, is he?"

"No. But he's part of my cluster. I owe him." Cheetah cluster for life, right?

"What's your world like? High or low RCP?"

"I think it's high?" you say. "It's not as bad as this, but it wasn't nice like the last place. The air was good. No storms or heavy wind. Clear. You could see—"

"A nine or so is considered high; two or three is utopia," Che interrupts. "That's where you get all your horrible shit under control and can keep a living world. The eggheads tag all the worlds we can access with various RCP levels. The nerds used to think the points of divergence would all be about national borders and great people. Like, suppose Hitler lived, or the Soviet Union collapsed. Shit like that. But the looming threat, the thread that runs through all these realities that makes the big changes that people like you and I give a shit about, is simply how the atmosphere and oceans were managed. Usually there's an accord. Paris, DC, they try to imitate the same thing they used to stop acid rain. Caps and trades. But sometimes they never even get to that point. Right now, this variation we're in, it's one of the worst scenarios."

You think about all the cool, breathable air in the world just before this one. "Then I'm definitely in a higher RCP world. People are scared of the runaway effect. A lot of people are bunkering underground, or in domes. I use a respirator outside. We're not using fuel anymore. It all got used up in the Collapse."

"Sounds like a shithole. But, you're a sister me. If you feel you need to go back, I can get you some weapons. You can try to find another Armand going the other way. But trips like that are fairly infrequent. I've been waiting to trap Armand here for almost a month."

"Can't *you* take me?" you ask. If she is doing favors and all. You feel somewhat paralyzed, because this is a situation so far outside of normal—how *can* you make a decision? This familiar face, however, knows more than you. You feel an instant desire for her help, her guidance.

Che shakes her head. "This truck's got one more jump in her, and we're going somewhere specific. And you can't take your buddy back there if you want him to live. You need to go forward."

"To a better world?" You're starting to get the hang of this. "Lower RCP?"

"The lowest," Che says. "You talk about bunkers and domes. This whole area was a massive dome once. This world ate itself alive. Kept putting leaders up that focused on chewing through resources, promising jobs over stewardship. And when the carbon from burning things filled the oceans, the pH meant huge die-offs. One collapse led to another."

"Sounds familiar," you mutter. And you think of that cool breeze on your skin and shiver. Something inside you almost aches when you consider trying to get back to what is familiar.

"But this world had resources, scientists working on keeping them alive as the air soured. When they cracked the veil between worlds, they found a garden world. An Eden. They built massive complexes to shuttle people over, thinking no one was on the other side. But they made a mistake. There were people, and those gardens were maintained carefully by people who had spent generations on wilding projects."

Che pulls out that same picture that the Shäd showed you. It's the clean, flattened-hair version of you, wearing a suit similar to Armand's. Che taps it.

"Which version of us is that?" you ask.

"This picture is a sister of ours that led her people from this high RCP world we're in right now to the garden world," Che says. "I killed her. That was my job. The garden world was my world, before I came over here to hide. It's where the invaders would never think to look for me. They think those of us from the garden world are soft."

But this sister of yours doesn't look soft, despite growing up in an Edenic world. She looks forged. Just like you have been. And even the sister in the picture looks forged. The leader of an invasion force that *had* to leave their dying world, or die themselves. Making hard choices.

"Those invaders to my world are the ones paying Armand anything he wants in order to get one of us, sister," Che says to you. "It's so that they can string us up in public for their people and have a good show as they execute us. Bread and circuses."

"Not true," Armand hisses. "The people I work for, they would take you in for training. They want insight; they need help to run the fight against these terrorists who have attacked us. They want to tap your genius."

Che shakes her head. "He's lying."

"And in exchange for your insight into fighting your terrorist double, they would make you richer than you could imagine," Armand says. "Join her, and you will be the enemy. An insurgent. On a list of enemies against humanity. And not just in the world she wants to go to, but any world that has trade with my people."

"No matter where he is born, he is always trash," Che says, and throws him into one of the cages. She locks it after him.

You stare at Armand for a long moment. "These better worlds. They can help Miko?"

"Yes."

"If I don't go to Armand's people, I'll be hunted. Because of something *you* did."

Che nods. "You'll be a criminal. A terrorist."

I look over at the three raiders who are done piling the goods up against the back of the trailer. "Why are you going over, now? What are you doing with all that?"

You're imagining that she'll tell you something about funding a revolution. How her people on the other side, invaded by the people from this world, need those things to fight their fight.

Che smiles. "They left so many here to keep suffering, once they got across, because the energy required to pierce through is nearly unimaginable. Each of those smuggler's trucks requires shareholders, backers, venture capital. Cross-world travel is rare. So I'm taking as many as I can get over. Count yourself lucky there's a spot for you."

You think about the treasures Armand has hoarded. They're all priceless things that the insanely rich obsess over. Portable cultural artifacts.

"I'll cross with you," you say, your voice breaking slightly as you realize this likely means you won't ever see your world again. But then, you've left it all behind before. Running across the desert in your bare feet, your hands covered in blood and hair hacked off, those manacles burning the skin of your wrists.

Five hundred miles away, your feet bloody, Cheetah cluster took you in and away.

Che pulls her mask back on, and you follow suit. The back door slides back down into a ramp, and there are hundreds of forms waiting in the dark-brown mist. Children.

People roughly shove the priceless art and precious metals off the back of the trailer onto the rocky ground.

Armand shouts from inside his cage, but the anger turns to coughing as the hot, acidic air roils into the trailer.

***

You've always gone forward.

This is the first time you've known what you're heading toward. An Edenic world without climate collapse. A world where invaders are fighting the people who lived there before them. Invaders who abandoned millions of their own to a dead world.

These starving refugees are packed into the back of the trailer or strapped onto the top with bottles of air. They are all taking the risk of death, or worse, when you all break through to that other universe. Che has been driving for several hours, hoping to get as far from core invader territory as she can, but air will be running out soon, and a decision to jump over has to be made shortly, or people will start dying.

You sit next to Che, shoved up together near the controls. Miko is by your feet. Refugees cram in everywhere inside the cab, some of them controlling the computer that Armand had, others just packed in to make the crossing. You'd watched the pile of pictures and all that glittered fall away behind the trailer in the mirrors. The old canvases curled up in the harsh air as you slid away. Acid rain began to drizzle on them.

"There are many more, on many different worlds, that need help," Che had said. "Even though their brothers invaded mine, I couldn't leave them back there to slowly choke to death. It wasn't their fault; it was their forefathers who did this to them."

You're never going to be a gunner again, chasing after convoys.

"Will it make a difference, a couple hundred?" I asked.

"We're just getting started," Che said. "I can save more."

But you wonder what all these new bodies will do to that Edenic new world? Will these descendants of a people who destroyed an entire world be able to make a change? Or would they understand how to treasure it, knowing how precious it was?

The world flips inside out, and you gasp.

When your vision comes back to you, you look up to see a forest choking the dirt road around the truck.

People rip off their masks as Che rolls the windows down. There are flowers. The distant chatter of animals. Scents on the breeze that fill the cab. A gentle wind.

She's taken you back to where this all started, geographically. But there is no toll road here. No farms. No dusty plains and electric cars with skulls and machine guns mounted on them. Just lush green and lungfuls of sweet air.

"Are you okay?" Che asks.

You wipe the tear from your cheek.

"This is what it could have been, where I come from," you whisper. "We could have done it."

Che stops the truck, as it's shuddering, and smoke has begun to leak out of the machinery in the front of the trailer. People are banging and shouting for her attention.

Afterward, she climbs on the hood as people gather before her.

"I give you my world," she says. "I ask only that you care for it as you would your child. Do that, and it will care for you as well. Now go."

The two of you sit for a while later near Miko.

"I'm sorry," she says. "I hoped he would make it. We could have helped."

You shake your head. "I've seen worse deaths." It was no worse a death than clashing with duty evaders on a toll road. He had never understood what was actually happening, though, and you feel bad about that.

When you stand, after burying him under the cool shade of a magnificent pine tree that makes you almost weep again, you move to Che's side.

"Take me with you."

"I don't know what my next move is. I snuck over to their world to see what it was like and hide from her death squads. I didn't really believe them when they said it was so bad. I didn't know so many were suffering. Armand and his ilk, they get financing to go and fetch the greatest minds from the remains of high RCP worlds. Or get 'servants' of famous people. But these are regular folk that needed saving, that no one wants. They deserve to live."

You nod.

You've been watching this version of yourself. And you've learned something. This isn't a confident, dangerous you like you first assumed when you saw her with the rocket launcher. No, this is a version of you that cares deeply. She put herself in front of the truck because she felt strongly.

She didn't know what would come next when she rescued these people. She just did it.

This version of you isn't calculating.

This version of you isn't looking for the next move.

This version of you came from ancestors who managed a planet successfully, not like your own failed ones.

You've been taught to take, to be strong, to run with a strong pack. Miko has taught you well. Having studied this Che for hours, you know you could break her. She hasn't spent a lifetime in the hot sand.

But the forests.

Her people let that be. And that is a different lesson. You need to follow her people. You need to help them protect what they've built.

And what you know, and have learned for an entire life, is survival. How to create fighters.

"You need to grow your army," you say to Che. "You need to bring more people over. More refugees from the other worlds. We can find more Armands to trade with, right, if there are many worlds and many copies of us in them?"

Che nods at you. "More death."

You smile crookedly. "And maybe more life. I think we bring things over to the other places that help them. And then we bring fighters, loyal fighters, from there to here. We can do this better. I can show you how to finish what you've started. Let me help you."

"It'll be dangerous." Che raises an eyebrow.

You know. But you know danger better than she does. All you've lived is danger.

Besides, is it better to be a king of a sandy hell, or a servant in a lush paradise? Her people had created something so special, you know that being one of their soldiers is the way to climb. You can taste ambition again.

This, you think as you move through sun-dappled forest, is a world to die for.

A world to fight for.

# Cost of Doing Business

*Nancy Kress*

Some people are always too late to the party.

Kayla moved around the edge of the crowd of protesters, the camera in her sunhat recording continuously, her tiny handheld controlling the zoom. Looking for the best shot, the one that would go viral and that her prose could make iconic. There: the old man sweating so much that the toddler on his shoulders had slid sideways, still clutching his little homemade sign: LEAVE OUR CLOUDS ALONE!

Yes. A better image than the protesters who had—pointlessly—chained themselves to the airfield fence. Better than the signs that required more background to understand—SAVE THE SAHEL and NO SOGGY MIRRORS—or the ones tied to wacko conspiracy theories: CHEM TRAILS TO CLOUDS TO CAPTIVITY. Better than the water vendors, making profit from heat prostration under the murderous Midwestern sun, hot as dragons' breath.

A woman in short shorts and crop top jostled Kayla. "Oh, sorry."

"No problem. But..." She made her voice soft and hesitant. "...isn't this just a one-shot thing, to help Chicago?"

"No. Don't you read anything? This is a fucking *test*. If it works, the government will go on to putting sulfates in the air for so-called global cooling, there'll be droughts in Africa and India, and the—" Her head swiveled to watch the planes take off from the heavily guarded airfield, their roar drowning out her words.

The planes, silver in the relentless sunlight, would seed the sky with aerosols that would cause the formation of water droplets, many more than would form naturally. The resulting cloud, a "soggy mirror," would drift east until it shaded Chicago, where the city's worst heat wave on record was shutting down businesses, melting asphalt, and killing—so far—269 people, most of them poor. The soggy mirror would reflect much more sunlight than normal clouds, relieving the worst of the heat caused by rising global temperatures, if only for a few days. Clouds did not last.

Neither would this cloud effort. Too expensive, too much public resistance. Did the protesters know that the decision to abort had just been officially announced? If so, they didn't believe it.

The crop-topped protester returned to educating Kayla. "Geoengineering is just playing God with the planet, and not only that—hey! Why aren't you sweating? You're wearing a c-suit! Luke, over here—another infiltrator!" She grabbed Kayla's arm, hard.

Kayla broke free and ran. Mistake, mistake, she shouldn't have worn the cooling-suit under her clothes, but she'd been afraid that otherwise she might faint in the 105-degree heat before she got the story. Running cranked up the automatic cooling of the suit, but no amount of cooling could give her the speed and agility she'd had twenty years ago. Luke was gaining on her left, and now two more men angled toward her from the right.

Two weeks ago, a journalist had been badly beaten while covering a fracking protest in Montana. All media was assumed to have a corporate soul.

Kayla stumbled and fell. She curled into a ball, covered her head with her hands, and waited for the blows. They would come—she'd seen the expression on Luke's face.

No blows. Instead, hands lifted her to her feet. "Ms. Serdenovna?"

Three men, one incongruously dressed in a dark suit; he must have had an incredible c-suit on underneath. The other two had shorts, tees, and assault rifles.

Kayla broke free. "Who are you?"

"I work for someone who would like to talk to you. We can, of course, ensure your safety." He gestured toward a long black car fifty yards away.

"I repeat, who are you? Who's in the car?"

"James C. Sullivan."

Kayla blinked. More protesters massed to her left. She saw the glint of sunlight on metal.

"Okay, let's go. But if that's not Sullivan..."

It was. Kayla recognized him; the whole world recognized him. Tall, austerely handsome, dressed more like a banker than a software geek, James Conway Sullivan had created a computer revolution, starting with his VR programs that turned an ordinary cell phone into a projector of room-filling, substantial-seeming holograms. Decorators and architects modeled room design; corporations created R&D presentations; history teachers filled classrooms with the Roman Forum; teenage boys holo'ed naked models in their dorm rooms. From there, more software doing more things no one had even realized needing doing, and then serious AI, until James Sullivan had become the richest man in the country, exceeding Bill Gates's fortune by half. Famously frugal, he spent almost none of it, except on security.

Kayla, like most liberals, despised him. People like him controlled 95 percent of the world's wealth, and this particular oligarch spent none of it helping the people who needed it most.

"Hello, Ms. Serdenovna. Please get in. I have a professional offer for you."

***

He took her to a Marriott with, blessedly, AC. The unpretentious suite had a small, well-equipped bar. Kayla, who had interviewed ambassadors and princes, could recognize a bespoke suit. Sullivan's wasn't.

He said, "Would you like a drink?"

"No. What is this offer?" Not that she was going to accept it. But she was curious.

"I'd like you to write a book about an endeavor I'm going to undertake."

"I'm not a ghost writer. Nor a business writer."

"Don't you think I know that? You write books about failures."

Despite herself, Kayla blinked at his bluntness. What an odd duck he was: a software genius with little online presence, a native West Virginian with the precise diction of a fussy English professor, a fantastically rich mogul who stayed at a Marriott. And what he'd said was true. Her first book, *Standing Rock*, had been about the collapse of the protest movement against oil pipelines. *Red Planet Denied*, which had won the Pulitzer, had detailed the failure of NASA's plans to put a settlement on Mars when all the agency's funding had gone, too late, into the political fires kindled from climate change.

"You hardly qualify as a failure."

"No. And neither will my endeavor. But I've followed your work for years, and you're the journalist I want to cover this."

"What is 'this'?"

"Before I tell you, you will need to sign a nondisclosure agreement. It commits you to nothing."

"I'm not interested, Mr. Sullivan."

He went on as if she hadn't spoken, but his gaze stayed locked on her. "You would have total and unfettered access for a minimum of a year to all my meetings, memos, and deals. There will be no restrictions on what you can write or where you can publish. The advance on the book is two million dollars, plus all expenses, of course."

Kayla felt her eyes widen; she'd never been a poker-faced negotiator. But she said nothing.

"The endeavor will be huge. Worth a year of your time, I promise."

Money was not going to lure her into being a flunky for James Sullivan. That gave her the luxury of being blunt in a way she never would have been with a source she actually wished to cultivate. "Tell me what the endeavor is. One sentence, before I sign any nondisclosure. You have my word that I will neither talk nor write about your one sentence."

Sullivan hesitated. Then he said, "All right. I'm going to tackle climate change by taking the United States off a fossil fuel-based economy."

She said, stupidly, "What?"

"I'm going to take—"

"I heard you. I just...Sullivan, the entire economy? You must know that's not possible."

"It is. Plans exist."

"I know plans exist! I wrote *Standing Rock*, remember? Our political and economic structures don't permit going green on anything but a pathetically small scale."

"They will."

She shook her head. "You're wrong. Also crazy."

"Then you can write another Pulitzer-winning book about my megalomaniacal failure."

Kayla stared at him. He was serious. "Why?"

"Because I won't fail. Sign the nondisclosure, Kayla. It commits you to nothing except another five minutes of blather."

This was true. She read the paper carefully, signed, and said, "I'll have that drink, after all. Two fingers of Scotch, neat. Okay. Talk."

He brought her the drink. "Corporations are all built on gas and oil, right? To manufacture, transport goods and people, market to consumers, invest and speculate and—"

"Don't talk down to me, Sullivan."

"Sorry. It won't happen again. So politicians back corporations to keep the economy going, and corporations back politicians. What is greater than that interlocking power structure?"

"Nothing," said Kayla, bitterly. Of the Standing Rock protesters, three had ended up dead and sixteen in prison.

"You're wrong. There is a force greater than corporations, at least in America. The vote."

Kayla laughed. Now she was sure he was crazy as well as egotistical. "Mobilizing the people? C'mon, Sullivan! You couldn't get voters to agree at gunpoint on a flavor of ice cream, let alone a plan to demolish the economy that provides them with jobs and cars and trips to Disneyland."

"That's the conventional wisdom, yes—that major paradigm shifts are impossible. And yet, once our economy ran on slavery, until it didn't. Once only white men who owned property could vote, until everyone could. Once women like you had to stay home and raise kids, until you had other choices. Paradigms do shift, have shifted, before. And I'll start small, with one model city."

"Oh, right. Stop climate change with a model city. Every 'model city' in history has failed. Building another mini-utopia in some desert is just stupid."

"Not a utopia, not in a desert, and not mini. This city will be different."

"Uh-huh." She sipped her Scotch. "And what is this 'model city'?"

"Flint, Michigan."

She couldn't help it; she erupted into guffaws, spraying Scotch all over the rug. "Flint! It's in bankruptcy yet again, it has an 11 percent unemployment rate, its power company shows huge losses, they still haven't replaced a third of those lead-contaminated pipes, and the FBI just named it the third most dangerous city in the country!"

"I know. Do you want to write the book?"

She still despised him; his ridiculous good intentions had not changed that. But this wasn't any ordinary project, and she was a journalist. Already ideas were rising in her mind like heat waves, shimmering with promise, coalescing into mirages of the book she could write. Failure and narcissism on an Ozymandias scale. Rise and fall of an oligarch.

She was tired of writing *New Yorker* articles but had no other book in mind. And Sullivan, to his credit, had not reacted with anger to her unprofessional insults. That boded well for a working relationship while she exposed his folly.

"Yes," she said. "I'm in."

***

She and Robert had dinner at their favorite bistro in Tribeca. Kayla had barely had time to catch a taxi from LaGuardia. Cars honked; drivers snarled at the usual Manhattan commuter traffic. She laid her head against the back of the seat and prepared to tell Robert that she would be traveling for the entire next year. It wasn't going to be an easy conversation, but not because of the travel.

They had met three years ago, to instant attraction. Despite the long separations of two high-powered careers in journalism, the attraction stayed strong. Kayla had never felt anything like it. When she won the Pulitzer, she had watched Robert carefully for signs of the professional jealousy that had wrecked her marriage ten years earlier. The signs didn't come. Robert was genuinely pleased for her, unthreatened, able to spend the next year singing backup to Kayla's literary hit. Their relationship was intellectually stimulating, physically exciting, emotionally satisfying. And yet, neither had ever brought up living together, let alone marriage. They remained a double-star system, just as strongly held to each other as they were separated by unseen centrifugal forces she didn't want to examine too closely.

"Kayla. I missed you. How did the veil protest go?"

"It fizzled. Geoengineering for climate control is a dead issue."

"You knew that already. Give me my kiss."

He was never self-conscious about public affection, something Kayla had had to get used to. They hugged, kissed, sat down, and ordered. After wine and dinner, when the mood was mellow, Kayla said, "I've got my next book project."

"Oh? What?"

"I'm going to follow James Sullivan—*the* James Sullivan—around for a year and detail this insane project he has, but right now I can only say—"

"James Sullivan? You can't be serious!"

"Just listen, Robert. It's not a how-I-became-rich book. He—"

"You're going to be a PR flack for Sullivan Enterprises? *You?*"

Kayla put down her spoon. Her crème brûlée no longer looked appealing. She said deliberately, "Of course not. This is something quite different. But I can't discuss it because I've signed a nondisclosure."

Robert grimaced. Kayla's belly tightened. A long, ugly silence spun itself out. Finally Robert said, "Sullivan has never given significant money to philanthropy—not ever. And he has unknown and untraceable amounts of money stashed in Cayman Island banks. Peter wrote that piece on him for *Forbes* and couldn't crack his people, and Peter's good. You can't know what you're getting into."

"Everybody at that level has money stashed in the Caymans. You know that. And I'm insulted, Robert, that you're not trusting me more on this. I know what I'm doing."

"Don't be insulted. I didn't mean that." He tried to smile, but it was so forced that Kayla had to look away. Which didn't help.

She made herself say, "What's next for you now that the *Atlantic* piece is done?"

"Baton Rouge. The *Atlantic* again. There's apparently a health crisis there, with rich suburbs like Centurion getting all kinds of help but the poor in the city getting none for the same problem."

This was why she'd first fallen in love with him: his deep commitment to writing about the huge disparity in how the climate crisis affected rich and poor. She said, "Heat deaths again?"

"No. Something much different—miscarriages. Way too many."

"Something in the water? Or buried and leaking waste under the McMansions?"

"Don't know yet. I leave tomorrow. Are you done with your dessert?"

"Yes." She didn't want the crème brûlée. She wanted Robert back, but his face was stony and his gaze didn't meet hers. She said, "Robert, this assignment with Sullivan isn't a PR job."

"It will be. He'll make it be."

"I said it wasn't."

He said nothing, but condemnation showed in every line of his face. Angry at his judgment, Kayla said coldly, "Have a good trip. Call me from Louisiana."

"I will."

Neither of them suggested spending the night together.

***

The press conference was held at the Marriott, which passed for a luxury hotel in Flint. Kayla, waiting in a small anteroom with Sullivan, had been told what he would say, and her mind still reeled. The ballroom filled with local and state politicians and major press. Secret meetings had gone on for the previous month, night and day. The governor and the mayor both worked the room, fending off questions while reiterating the announcement that had brought everyone here, which boiled down to, "Something big is going to happen in Flint, and it involves James Sullivan."

After weeks with Sullivan, Kayla still couldn't get a bead on him. Completely open with her about his business transactions, he never said anything about a personal life, and none of her interviewing tactics had coaxed him into even the simplest revelation. Nor had her private digging yielded more than everyone knew: born in Florida to a postal worker and a housewife, now both dead. No siblings. BS in computer science from Stanford, where he seemed to have made many contacts but no close friends. Never married. A few girlfriends, none serious, all of whom said he was kind and reasonably generous but very secretive. "It's like I never got into his mind, you know?" said one. "And he worked all the time."

His employees in Silicon Valley said they seldom saw him; he left the day-to-day operations of Sullivan Enterprises to his COO, Marcus Calder. Probably wise—Sullivan's buttoned suits and stiff formality were a bad fit with a software company's style, which was given to T-shirts, arguments about zombie invasions, and friendly insults among programmers. Calder met with Kayla only because Sullivan told him to, but the meeting was frustratingly unproductive; Calder guarded whatever his boss's secrets might be like a dragon guarding treasure.

The Marriott had erected a makeshift dais in the ballroom. The governor, mayor, and other bigwigs climbed to their seats on the dais. Sullivan walked out and sat on the far left. Cameras were turned on; security stood by. The newspeople who had already ferreted out pieces of the story looked tense and, for newspeople, openly fascinated.

The governor made a welcoming speech. Kayla, who despised him for a self-serving idiot, rolled her eyes. Then the mayor took the podium.

"It's my great pleasure to make two major announcements about our city. First, as you already know, Flint has been in a public health state of emergency for too many years. Budgetary concerns have made it impossible to replace as fast as we would like the water mains delivering unsafe water."

Unseen, Kayla snorted. *Too many years* was an understatement. A third of the pipes were still contaminated with lead, lawsuits drowned the city, three officials had gone to jail, and Flint was once more in receivership.

"However, all the remaining pipes will be replaced immediately, by which I mean starting tomorrow, both providing safe water to Flint and creating hundreds of good-paying jobs. This is possible through a five-hundred-million-dollar grant from James Sullivan, who has bought and equipped the old Johnson building to hire workers, train them, and begin pipe replacement."

The room exploded. The mayor raised her hand for quiet—and were her eyes wet? No, just shining under the lights...weren't they? But there was no mistaking her grin.

"The second announcement is that the city will be switching its main energy company from Consumer Energy, which has served us well, to a new company with lower rates, Flint Energy Consortium. Everyone in Flint should see a drop in their energy bills next month and going forward."

"How?" Kayla had asked Sullivan when he'd first told her this.

"I negotiated a deal with CME, the parent company, to buy energy from them and sell it back at a reduced rate to the city. Flint wasn't very profitable for them—too much poverty, too many businesses closed."

"And you're making up the difference from your own fortune."

"It will pay off, eventually."

"No, it won't." Kayla said. He liked her to push back against his statements, as in their first meeting, and she usually learned more when she did. "You won't see any profit. And you're still using energy from fossil fuels."

"Not for long. Already the Consortium is buying fuel mainly from Michigan's wind farms and hydroelectric plants. I'll build more wind farms, install solar arrays with the newest capture-and-store features on public buildings."

"What will all this cost you?"

Sullivan thrust his hands deep in his pockets, looking for a moment almost boyish. Then Kayla realized the deliberateness of the gesture. She hadn't known he was capable of humor. He said, "It'll cost a lot. I'll give you figures later. To take back a city, you have to start with its energy sources, and energy companies only sell either when they can make a profit or when they're forced by law. I don't make laws."

"Is that where you're going? Buying a city so you can run a private dictatorship?"

"No. Really, Kayla, I know you better than that. You will have done your research. This mayor is nobody's tool."

True. In a corrupt state, Leah Goldman had proved incorruptible. Kayla said, "Flint will cost you billions."

"I have billions."

"And have never given as much as a dollar of it to charity. So—why?"

Sullivan shook his head. "I already told you. You don't believe me, which is fine. It's why I chose you."

Kayla said nothing. The look on Sullivan's face had baffled her. If she'd had to give a name to it, she would have called it sadness.

The press conference made headlines across the nation, professing equal parts wonder and skepticism. But Sullivan did exactly what he said. By the next day, people hungry for work stretched in a line around the block for jobs laying pipe. When the salary was announced, the lines doubled, then tripled.

***

She called Robert, whose light-created holo-presence irritated her by saying, "When does the other shoe drop?"

"Are you sure it will?"

"Of course it will. Converting you to rule by the 1 percent, is he?" And then, when he saw that she was too angry to even reply, he said, "Sorry. I'm stressed out. You can't believe how hot it is down here."

Of course she could; she'd been in Louisiana in late August. But fear kept her silent. Was she losing him? Looking at his tired face, she knew she didn't want that to happen. But neither was she willing to lose herself.

"Robert, how is your research going?"

"Slowly. There's a clear pattern of early-pregnancy miscarriages, and it's a statistically significant percentage. The CDC is here and the EPA and all the rest of the alphabet, plus an army of biologists and virologists. So far nobody knows anything. No fetus has been further along than four or five weeks, but the personal stories I'm collecting are heartbreaking, even for a curmudgeon like me who never wanted kids."

Kayla had never wanted them either, but that was irrelevant. She said, "When are you returning to New York?"

"Not till I get answers. You?"

"Ditto."

Then there seemed very little to say. Heartbreak hung in the silences.

***

Six months later, the Flint water crisis, which had dragged on for years, was over.

Heating and lighting bills dropped by 20 percent.

When the pipe replacement finished in Flint, workers were retrained to build a new utility wind turbine at a cost of two million dollars plus labor. More people were hired to build a five-billion-dollar light-rail system. Sullivan worked closely with the Mass Transportation Authority. Retail, restaurants, and repair companies expanded to serve the new workers. By November, when midterm elections were held, unemployment in Flint had dropped five full percentage points. The mayor was reelected in the largest landslide in Flint history. Sullivan was negotiating with GM for electric vehicles for city employees.

The mayor appointed Sullivan as ombudsman, a city position charged with representing the interests of the public.

Kayla walked around town, interviewing people, taking photos of the new flowerbeds, playgrounds, shops. She ate lunch at Angie's Diner, eavesdropping on construction workers. When she returned to the hotel, Sullivan had bought the *Flint Journal*.

The newspaper, which dated from 1876, had gone from a daily paper—Flint's only one—to publishing solely on weekends. Within a month, Sullivan restored daily circulation, hired more staff, ran a lot of articles on the climate crisis. He also bought an abandoned factory that had once made parts for the auto industry and had it retooled to produce solar panels, which he subsidized. Heating and lighting with solar became more affordable than drawing from the grid.

As they prepared to leave Flint, Kayla realized that she didn't know Sullivan any better, but she did like him more. He worked sixteen hours a day and then disappeared into his hotel room, but during meals and on the way to and from meetings, she'd discovered that he had a dry sense of humor and, incongruously, enjoyed the poetry of Robert W. Service. Kayla burst out laughing. "No, really? 'The Shooting of Dan McGrew'? I thought you'd have better taste!"

"No," he said, smiling, "you didn't."

That was the most intimate conversation they'd had. Worse, her conversations with Robert weren't much more intimate than those with Sullivan. Mostly they talked about the miscarriage crisis in Louisiana.

It was evident early on that the pregnant mothers showed liver changes that didn't affect them but disastrously affected fetuses under two months, leading to spontaneous abortion. A virus was suspected, but no one could find it.

"Viruses can be notoriously hard to isolate," Robert said during an early-morning holo-call. He looked tired but also wary. He still thought less of her for working with Sullivan, she still resented his condescension, and the digital air between them felt thick and painful.

"What does the epidemiological map show?"

"Distinct clusters and lines of transmission but nothing that can be pinned to a source, or to a maternal profile, or to anything at all."

"What about the fetuses?"

"They don't show any single pattern. Kayla, all this is in the news nearly every day, so why are you quizzing me?"

"Because we're supposed to be in love, remember?"

"Supposed to? Does that mean you're not?"

"I am. But I don't think you are, anymore."

Long silence. Then she knew. "You met someone else. In Louisiana."

"Yes."

"Who is she?"

"A CDC pathologist. I was waiting to tell you until I knew that we—"

"Which 'we'—you and me or you and her? I suppose she has absolutely pure ideals, unlike corrupted me?"

Robert ran his hand through his hair. It was thinning. "Kayla, I... I'm sorry."

"Me too," she said and cut the connection.

In her hotel bedroom, she sank onto the floor, back against the wall, and wrapped her arms around her body. The room felt cold. She rocked back and forth, trying to soothe herself, forbidding herself to cry. Kayla Serdenovna did not cry. But it was a long time before she could make herself get up off the floor, could breathe normally around the hard ball of pain in her throat, could get dressed and go down to the lobby to meet Sullivan for the day's work.

He noticed instantly. "Are you all right?"

"Yes. Let's go. Don't look at me like I'm dying, James—I'm not."

"Well, good," he said and treated her no differently than before. But she caught him looking at her now and then with concern.

***

A year after his campaign began, the crime and unemployment rates in Flint had dropped precipitously. Roof gardens, the plants free to anyone who wanted them, had begun to send leafy trailers spiraling down old walls. The city council voted to divest itself of all stocks connected with oil and gas extraction and to reinvest the money in renewables. Press gathered regularly to examine and write about "the Midwest miracle." Not all the articles were complimentary. "Paternalism at Its Worst." "Sullivan Buys Himself a City-State." "The Prince of Flint."

Kayla said, "Flint cost you about twenty-five billion dollars."

James only smiled.

"That's maybe one-sixth of your fortune, and it's one city. Even if you spend the rest to transform five more smallish cities, so what? You're broke, the fossil-fuel companies are still merrily drilling and fracking and plumbing tar sands, and Earth is warming more than ever. So what have you really accomplished?"

James sipped his Scotch. He was drinking more, but he never seemed affected by the liquor. They sat in an airport bar, luggage already checked for New York, waiting for their flight to be called. Sullivan Enterprises owned a plane, but his executives used it more than he did. Flying commercial when you were the richest man in the country seemed affected to Kayla; whatever James was, he was not a man of the people.

Still, she had extended her contract with him for another year. This whole insane project now absorbed her. Digging had produced a few more tentative leads, which she was determined to follow.

Finally he answered her. "If you want people to say no to fossil fuels—enough people to matter, not just the liberals who are already going solar and protesting pipelines and planting trees in suburbia—you need to show them something to say yes to. Flint is artificially created by my money, yes, but it's still a model of how a city can look. And it's just the start."

"But the start of what, exactly?"

"Kayla, do you know how much it costs to be elected president? Less than a billion dollars. And no, I don't want to run for president. Nor senator nor congressman."

"Yet you're putting together what looks a lot like a presidential-level, ground campaign network in every major city in the country. I'm supposed to have access to everything you do."

"You have it."

"To the 'what.' But not to the 'why.'"

"You already know the why. I'm going to take the United States off fossil fuels."

The wall again. She could get only so far before she hit it. "James—"

"No pushback right now, if you don't mind. They're calling our flight."

***

By June, $CO_2$ in the atmosphere had risen to 500 parts per million. Greenhouse emissions in the United States had surged again, after several years of falling. The decreases, Kayla knew, had been due to moving factories overseas and emitting there. The global average temperature had risen another tenth of a degree, edging up to the three-degree increase that climatologists now called critical. Earth could become a planet without winter, or at least without snow. Certainly it was hot enough in Oklahoma. Sweat dripped into Kayla's eyes despite the c-suit under her loose dress, but at this protest, no one noticed her.

Just outside the fence surrounding the big, ugly machinery, protesters chanted and screamed and waved signs. Fracking caused 30 percent more methane emissions than conventional extraction, but the chants and signs focused not on data but on emotion.

"No more earthquakes!"

SAVE MOTHER EARTH!

"No more earthquakes!"

FRACKING IS FATAL.

"No more earthquakes!"

LEAVE THE OIL IN THE SOIL.

The last had more passion than accuracy; this fracking site was after natural gas, not oil. Kayla moved closer, angling for a good shot. Overhead, surveillance drones added to the noise. Not, Kayla noted, many press drones, and those she did see were only from local media.

And then came another sound.

*Oh, Christ, no.*

Two military Strykers rolled up the road, followed by trucks carrying National Guard. Soldiers in riot gear, tear-gas canisters at their belts, leaped down and moved into formation around the crowd. Someone with an PA system blared, "You have the right to peaceful protest, but do not attempt to enter private property beyond the fence. Repeat, do not attempt to enter private property beyond the fence."

The crowd suddenly buzzed like the world's largest wasp's nest.

It hadn't looked to Kayla like anyone had intended to climb the chain-link fence. But dared like that, the inevitable young man did.

Early twenties, floppy hair falling over his forehead, he climbed the chain-link and turned defiantly, a little wobbly, to face the soldiers. In the sudden silence he screamed, "Go away, and take this fucking gas company with you!"

*No, no, don't do it, kid…*

She had seen such protesters, had written about them, before. In Greece, in Romania, in Canada, in Ecuador, in the United States. Sometimes the protesters won—all of France had banned fracking—but more often they did not. In China, environmental activists had been shot. In Nigeria, they had been hanged, burned out, mowed down with rifle fire. Milder militarized responses had resulted in people choking on tear gas, crying out with injuries, under arrest. But the fracking, drilling, tar-sand extraction, pipelines, and oil-train wrecks all continued.

The young man swaying on the fence, lifted one sneakered foot, and set it down on the other side.

Soldiers raised their rifles.

Then three large men rushed forward and wrestled the young man off the fence. He screamed; they screamed; protesters milled around. The air still felt charged, but Kayla knew there would be no violence, not here today. Fellow protesters carried away the still-struggling young man, his sneakers vainly trying to kick whomever he could.

Kayla's cell beeped in the pattern that meant breaking news. It referenced a *New York Times* article; she opened and read it.

Robert's friend and rival had beaten Robert to his scoop.

*Virus Causing Miscarriages Identified*
*CDC Warns of "Shocking Implications"*

*A Special Report by Peter Armbruster*

*Today the Centers for Disease Control and Prevention in Atlanta announced at a press conference that scientists have finally isolated the virus causing early-pregnancy miscarriages in four southern states. The virus was identified through a joint effort by the CDC, the United States Army Medical Research Institute for Infectious Diseases, and the US Public Health*

*Service, said lead epidemiologist Dr. Jane Schilling. "This virus has shocking implications," Schilling said. "We never expected this."*

She went on to explain that the miscarriages in Louisiana, Texas, Alabama, and Mississippi were caused by a virus temporarily named PV-45, which seems to be "a spontaneous mutation of a common virus."

*PV-45 interferes with the way the human liver processes the chemical toluene, also called methylbenzene. Toluene is a component of both gasoline and jet fuel.*

*In most PV-45-infected people, the liver simply figures out other ways to dispose of the temporary toluene overload. For pregnant women who are infected, however, the situation is different. Toluene readily crosses the placenta and causes miscarriage. At least 900 pregnancies so far have been ended by PV-45.*

*Oh, shit*, Kayla thought. That first reaction was followed by two others: Robert would be disappointed that Peter, with his connections, got this first. And then: James had just gotten a huge, unearned gift for his crusade.

Immediately she felt ashamed of herself. This was not about either Robert or James.

*It has been known for decades that fetuses metabolize toluene differently from adults. In low concentrations, exposure to toluene does not affect pregnancies. High exposure, especially if it is chronic, can cause fetal gasoline syndrome, with results similar to fetal alcohol syndrome: low birth weight, facial abnormalities, and brain damage. However, PV-45 seems not to lead to fetal gasoline syndrome, common among chronic "glue sniffers" trying for a euphoric high, but rather to spontaneous early miscarriage.*

*Scientists have not yet identified how the virus is spread. "It doesn't seem to be airborne," Schilling said, "but we need much more research on both its transmission vector and its effects. Most of what we know about methylbenzene effects comes from animal studies, which may or may not be applicable to humans. We do know that the fetal, larval, and egg stages of all kinds of animals are more susceptible to toxins than the adults—oysters, sea turtles, dolphins. Still, PV-45 is uncharted territory. All sorts of new diseases have emerged, or mutated, in the last twenty years that we've never seen before."*

*How to Protect Yourself*

*Who is at risk for PV-45? So far, the known miscarriages have clustered along heavily trafficked highways and among workers in factories that use toluene in manufacturing, such as printing, adhesives, and footwear. There has also been a significant increase in miscarriages downwind from two Louisiana oil drilling sites.*

*The CDC recommends that pregnant women, and those who think they might be pregnant, avoid gasoline, diesel fuel, and gasoline products, including paint, glue, and any household cleaners containing toluene. The OSHA safety threshold for toluene is 100 parts per million, and refueling your car can expose you to twice that amount. So, until more is known, let someone else pump gas for your car. Postpone painting in homes and workplaces. Avoid areas of heavy traffic. Don't—*

*Avoid areas of heavy traffic. Yeah, right. Tell that to all the pregnant women living in houses bordering interstates. Or standing at bus stops in crowded cities, breathing exhaust, to get to the jobs that keep their families fed.* Peter had always been an elitist.

The fracking protest had lost energy. Some demonstrators left in fear of the military presence. A few had succumbed to the heat and were being helped away by friends or relatives. None of the remainder had yet seen the *Times* article about PV-45.

*Toluene.* Such a pretty word. It sounded like a flower, or a girl's name.

Kayla returned to the article. Her phone rang. James.

"My God, Kayla, did you see—"

"I saw it. All those poor women hoping to be mothers…but, James, this changes everything. And in your favor."

"No, damn it—don't you see? This makes it worse than ever for what I want to do."

She was bewildered—that didn't make sense. What was he seeing that she was not?

Then she saw it.

"Meet me in DC," he said. "We have a lot of work to do."

***

Powerful corporations are powerful. A tautology, of course, but so easily forgotten that Kayla wanted to have it tattooed on the inside of her eyelids.

It took the fossil-fuel PR machine less than twelve hours to come up with "proof" that PV-45 was somehow created, or spread, or at least being capitalized on (it varied by corporation) by environmentalists intent on baselessly attacking gas, oil, and the American way of life. Shell, BP, and Exxon pledged billions of dollars to fight the virus.

The stock market plunged, throwing the global economy into confusion and dread.

The media raced to interview scientists, pregnant women, corporate executives, politicians, and anyone else with an opinion, which was pretty much everyone. Tearful women who were pregnant, or hoped to be, fled north from the infected states, usually to friends or relatives but some to a hastily opened "Pre-born Refuge" in Nevada, run by a religious group. The president gave a press conference each day, urging calm. Nobody listened. Kayla, outraged, hoped the crisis would stop with interviews and press conferences, but she knew better.

Protesters against the "dirty tricks" of environmentalists torched Sierra Club headquarters in Oakland.

Protesters against auto emissions blocked and closed down Interstate 65, backing up traffic and delaying delivery of goods.

Thousands refused to drive their cars, and tens of thousands who did so anyway were criticized and sometimes attacked.

Several countries put a temporary ban on travelers from the United States until more was known about the virus.

By the end of the week, PV-45 had spread to Florida and Georgia.

Polls showed that about a third of the country believed that the virus was indeed created by environmentalists to push "green fascism." It was the same charge that had been made when Antarctic ice shelves had collapsed: "They blew it up to make us think global warming is real!" By now, nobody denied global warming was real, but anti-environmentalists still believed it was (pick one) a normal fluctuation, not as bad as the hysterical left said, or soon to be fixed by a technological miracle that was going to come along any day now.

Another third considered the virus an act of foreign terrorism that should lead to bombing someone, somewhere.

Slightly less than a third thought PV-45 had evolved naturally but that the fossil-fuel companies should do something immediately; they did not agree on what.

The remainder attributed PV-45 to, variously, China, the End Times, a mad scientist somewhere, or an international cabal of illuminati, Jews, or zero-population fanatics.

James spent the week in DC meeting with lobbyists and members of Congress, with several side trips to various corporations. Kayla went on the first few of these, which were not very productive. He suggested that she cover the CDC instead.

"Stuck in a mob of reporters penned up somewhere in Atlanta until something breaks? James, this isn't what you promised me. I go where you go."

"I know, Kayla, but nobody foresaw this, did they? I have corporate crises of my own to attend to in Silicon Valley, and they're proprietary to my company and not connected with your book. The stock of Sullivan Enterprises is taking the same dive as everybody else's. You know that."

She did.

He added, "I know Francis LaHaye; we were at Harvard together. I'll get you a private interview. It'll be only fifteen minutes, if that, but it'll be an exclusive."

Nobody, not even Peter, had gotten an exclusive with the beleaguered head of the CDC. Kayla went to Atlanta.

She was there when the CDC announced that it had found the transmission vector for PV-45. Like so many other diseases that had once been tropical but had moved north with global warming, PV-45 was carried by a mosquito. *Aedes albopictus*, which unlike most mosquitoes, fed not only at dawn and dusk but during all twenty-four hours.

***

"Kayla," said the voice on her phone, and she froze in the act of brushing her teeth after a long day of interviewing, writing frantically, and checking the news every fifteen minutes. The air in her hotel room suddenly felt as hot as the stifling city outside. Then the room went cold.

"Kayla?"

"Hello, Robert." She mastered herself, but not her voice, which she cursed for its shakiness. They hadn't spoken since the breakup. She had deliberately not followed his career or his life.

"Look, Kayla, I'm not going to pretend this call isn't awkward as hell. Your phone says you're in Atlanta; is that right?"

"Yes." She fumbled with the tracking on her own phone; he was calling from Minneapolis. Why?

"Are you still working with Sullivan? If you're not, then this call is pointless, and I won't bother you. But if you are, there's something I think you should know."

"I am still researching James, yes." There, that tone was better—neutral, professional.

"Okay. I have sources all over—you know that—and one involved with Sullivan's ground campaign for—you knew he was building an extensive organization, right?"

"Of course I did. Don't insult me." James had been at his massive anti-fossil-fuel campaign for a year; it rolled out on Sunday. Kayla knew even more about it than James thought she did.

"Sorry," Robert said. "But Sullivan's campaign isn't what I'm calling about. I can't tell you who my source is for this, or anything about him or her, but the source has high-level contacts in the CDC and told me that it isn't releasing everything it knows about PV-45. The source doesn't know what's being held back but says it's big and wanted me to investigate. I can't, but through Sullivan, you probably have access to people that I don't."

Kayla still held her toothbrush. She rinsed it and put it down, letting the ordinary little action focus her. Robert sounded angry and aggrieved. At not having the access that, he rightly assumed, she had? Or at James, whom he'd always distrusted the way he distrusted all corporate brass?

"How reliable is your source?"

"Very."

"What else did the source tell you?"

"Nothing. I know it's not much."

"No, it's not. In fact, it's nothing. Just a rumor."

"No. This person is completely reliable."

"Even if that's so, why are you telling me? The CDC isn't connected with James. Why not tell Peter?"

Now he sounded angry. "I thought you should have it because you're a better journalist than Peter. Sorry if my tip offended you."

"It doesn't. But you sound offended. Robert, why are you in Minneapolis? You hate the Midwest."

"Jenna has family here. It's safer. She's pregnant."

*Put out one hand; steady yourself on the wall. Walk to the chair. Sit down.*

"Kayla? You still there?"

"Yes."

"I know what you're thinking. I did always say I didn't want kids. But with Jenna...well, it's different now."

Different than it had been with her. And Kayla knew, because she knew Robert, that he would only have said that if he believed that she was as completely over him as he was over her. He did believe that. She needed him to believe that.

She needed to believe that.

From long journalistic experience, she found the right tone. "Congratulations. But now that we know PV-45 is spread by mosquitoes, it will be easier to control it. We eradicated malaria-carrying mosquitoes from the US. We can do this, too."

"Yes, I think so."

"Thanks for the call."

"Sure. But will you call me if you find out anything more about the CDC?"

"Yes." *Have to keep Jenna and her baby safe.* "I'll talk to James as soon as he returns from Silicon Valley."

A long silence. "Sullivan isn't in Silicon Valley. He's in Miami. It was just on the news."

"Yes, of course. Silicon Valley was before Miami. Good night, Robert."

***

"You lied to me."

James sat in the living room of their shared hotel suite in New York, holding his volume of Robert W. Service poems. His gaze didn't flinch from Kayla's. He waited.

"You weren't in Silicon Valley; you were in Miami. Why? And why lie about it?"

"Because my reason for going to Miami was completely personal. Someone I had to see there. A woman, all right? It didn't concern you."

He had never been rude to her before. Rudeness didn't disconcert her but pain did. He—James Sullivan!—looked near tears.

"I'm sorry," she said awkwardly. "But you should have told me. I wouldn't have asked personal questions."

He looked at her and laughed. An unhappy laugh, and mocking. And he was right. She would have asked personal questions.

James said, "I don't want you to try to find her or to talk to her. My personal life wasn't part of our deal. Nor do I want to discuss this with you. I'm going to get some sleep—the campaign rolls out tomorrow."

After he'd gone to his room, Kayla picked up his book. He'd marked a passage of Service's wretched doggerel:

*Go couch you childwise in the grass,*
*Believing it's some jungle strange,*
*Where mighty monsters peer and pass,*

*Where beetles roam and spiders range.*

*'Mid gloom and gleam of leaf and blade,*
*What dragons rasp their painted wings!*

"Huh," Kayla said. She hadn't promised not to find the woman in Florida. Nor had she told James that, in her opinion, the campaign would do no good at all.

***

It didn't.

Not for lack of planning. James had spent billions. An army of canvassers handed out leaflets. TV and radio spots played as incessantly as if this were a presidential campaign. Online ads, blogs, tweets, videos, and pop-ups cluttered every screen in the country. Yard signs appeared. Everything, the simple and the complex, had one message: *Eliminate gas and oil and you slow global warming*. Nobody cared. The country was focused on eliminating mosquitoes.

By 1951, *Anopheles*, the mosquito that carried malaria, had been eradicated from the United States. All the old weapons were hauled out: insecticides, repellents, netting, draining breeding grounds. Even an old cartoon character, Annie Awful, an *Anopheles* dripping blood with the original voice-over: "Her business is robbery and coldblooded murder. They call her Annie Awful. She's a thief and a killer. She stops at nothing." Now Annie—and never mind that *Aedes* rather than *Anopheles* was the carrier this time around—was projected as a truly scary hologram running on all Sullivan Enterprises software. Handheld mobile labs made spot diagnoses of everyone possibly infected while also tracking and updating epidemiological data.

The miscarriages increased.

Kayla, dashing from city to city to cover everything she could, talked to James by cell several times a day. "We're changing the materials," he said. "The new message is: no gas and oil, no miscarriages. The new spots and leaflets and all aren't as polished as the old stuff—no time. But it's all we have."

It wasn't all he had. While Kayla concentrated on information for her book, James had what seemed to be an army of journalists covering everything else he was orchestrating. Technically, they all held press credentials from the *Flint Journal*, which gave that small paper more coverage than NBC. James told her the major developments in time for her to get to them, but not even he foresaw the CDC's next press conference.

"The PV-45 virology team," announced an exhausted Francis LaHaye, "has evidence that although the generation of mosquitoes carrying PV-45 has been largely eliminated from the six infected southern states, the next generation of mosquitoes is also carrying the virus. That's why miscarriages have increased, not decreased. The best explanation for this is that PV-45 has an alternate host. In Africa, for instance, malaria remains a scourge because infected mosquitoes bite cattle, who then carry the malaria parasite in their blood, and when a new generation of mosquitoes bites the cattle, they become infected and then bite people. That seems to be happening here."

A reporter at the press conference called out, "What is the other host?"

"We don't know yet."

Three weeks later, the CDC announced the alternate host: the brown rat, *Rattus norvegicus*, the most common rat in the world and the most successful mammal on the planet, after humans. No place on Earth had succeeded in eliminating the brown rat.

Miscarriages continued to increase. But PV-45 and its twin carriers were, as James's relentless campaign insisted through airwaves, print, and pixels, only half of the equation. PV-45 might be impossible to eradicate, but it caused miscarriages only because it acted on toluene breathed in by pregnant women.

No gas and oil, no miscarriages.

***

"Ain't gonna happen," Kayla said to James.

"Not completely, no. You need fossil fuels for some manufacturing, some products. But they don't have to be completely eliminated.

Just mostly. That would slow global warming enough to make a difference."

"Even if it did—even if, which I doubt—the US isn't the only country emitting greenhouse gases. Or even the major one."

"No."

"The entire global economy runs on fossil fuels."

"It doesn't have to."

"James—"

"I'm tired, Kayla. Leave me alone for just tonight, please."

His usual upright posture had given way to slumping in his chair, whiskey in hand. The whiskey was his third, very unlike him, even after their long day of brutal meetings with K Street lobbyists in DC. Kayla took the opposite chair, torn between compassion and journalism. A tipsy source often let down his guard to talk more freely.

She looked more closely. James had lost a lot of weight; his dark-gray suit hung on him. He seemed not only exhausted but eaten away from within. Kayla had a sudden horrified thought. "James, are you ill? Seriously, I mean?"

He smiled, the whimsical grin that so surprisingly changed his entire face. "What you mean is: do I look like this because I'm dying, and is imminent death the real reason for my sudden philanthropic crusade? No. I don't have cancer or anything else fatal."

She took a chance. "Then it wasn't a doctor you went to see in the Cayman Islands last week, without stopping in Miami to visit a woman or anyone else?"

Long silence. The grin had disappeared.

"You're good, Kayla. You must have people working with you that I don't know about. And you probably think that I have illegal funds in the Caymans."

She waited for him to confirm or deny it. He did neither.

"Goodnight, Kayla."

"James, wait a—"

"Goodnight."

She sat a while longer, thinking, but no matter how she shifted the pieces, they didn't quite fit together. A campaign to stop climate change, waged single-handedly by a fundamentally decent, obscenely rich man who would not be at all rich at campaign's end. A secretive nature that wanted Kayla to write a book about him. Illegal and untraceable offshore money. That deep, inexplicable sadness. The lie about a woman in Miami; Kayla's digging, with help both hired and as favors, had 90 percent convinced her that it was a lie. Journalistic leads that, time after time, ran smack up against stony silence.

That night, she dreamed of the grandmother who had raised her, dead for ten years now. Grandma Ann had not been a warm woman, or a particularly intelligent one, and Kayla had fought with her for decades, fights that had left both of them exhausted and unhappy. "Your generation don't know what it's like to sacrifice for the common good," Grandma Ann had often said. "In the war we all did without, ration and ration alike, and nobody died of not having chocolate or nylons or gas for the car. Spoiled, all you kids. And the next generation even more."

*Shut up, Grandma Ann. Let me sleep.*

Wasn't gonna happen. At least, not tonight.

***

By September, things started to turn. Slowly, and at first in small ways.

The Rhode Island state legislature passed a law banning all natural gas extraction and stating that nature had an "inalienable right" to be protected. Such a law wasn't new—the city of Pittsburgh had one since 2010—and Rhode Island was hardly a major fracking site anyway.

New Mexico was. The legislature passed a law banning all gas and oil extraction. PV-45 was moving north and west.

An industry consortium immediately filed a lawsuit against the legislature, arguing unfair restraint of trade.

"All right," James said. "This is where it happens."

"Or doesn't," Kayla said. "This is your big gamble, isn't it? I want the latest cost figures."

He transmitted pages and pages of data, already prepared for her. She went over them sitting up in bed in yet another nondescript hotel room and knew that she was looking at the biggest gamble any billionaire had ever made with any fortune. James was spending just about everything he had on multiple political campaigns, none of them his.

Over a year ago, he had said to her that only one force in the United States was greater than corporations: voters. Now, to an extent that she had not imagined even him capable of, he was doing what in Kayla's experience people seldom did. He was putting his money where his mouth was.

All his new candidates for national election in the November midterms were eco-green. No, not just green: bright, dazzling, Irish-countryside emerald. Those up for re-election were environmentalists who'd had to restrain their actions under the previous administration. They would not do so now. James had precision-targeted where to put his campaign money. If even two-thirds of these senators and congresspeople were elected, Congress could enact sweeping environmental reforms.

For the first time, Kayla thought, *My God, he might just do this.* It wouldn't have happened without PV-45, but PV-45 had happened. Did—

No. Insane thought. *Rein it in, Kayla.* She was a journalist, not a fantasist.

The candidates began, or ramped up, campaigning the next day. Intense, expensive campaigning. Journalists discovered who was financing the campaigns, and stories exploded across airwaves, pixels, print. Kayla flew to Colorado, that perpetual political battleground, and interviewed random people.

*"Hell, yes, I'm voting for Louis Tallerton. It's about time we stopped destroying the planet we all have to live on."*

*"Fuck, no—not Tallerton or any other green fascist. You know how much our jobs depend on gas and oil? No breadlines for my family!"*

*"He can't win—grow up, girlie."*

*"We need change. This is, like, a crisis!"*

*"I don't think...the issue is...who is Tallerton, again?"*

*"That TV ad said..."*

*"That other TV ad said..."*

*"I read online..."*

*"It was on Facebook that..."*

A hundred street mini-interviews, and Kayla remembered only one vividly. The woman gazed at her for a long moment. Then she pulled from her purse a worn wallet, and from the wallet a picture of a young woman smiling shyly into the camera. The woman, maybe thirty, had medium-brown skin, hair in dreads, and a smile like the sun.

"This is Mary, my daughter. She and her husband been trying eight years for a baby. Being a mother is all my Mary ever wanted. Then she got pregnant. Got a little ultrasound picture and sent it to me on the email. Then she lost the baby. Mary and Darryl, they live right by Highway 85 in Montgomery. Her heart's broken, and mine, too. Yeah, I'm voting for Tallerton. I don't usually bother with no elections except for president, but I registered yesterday."

There were a lot like Mary's mother.

A candidate for Congress in Colorado was attacked at her own rally. She asked for sacrifices to be made by the voters: "just as the gas and oil companies are making sacrifices by paying for new technology to remove carbon from the air." The crowd howled—actually howled—and mobbed the stage. Only a double line of cops got her away safely.

James said to Kayla, "Politicians who ask for sacrifice don't get elected. Not unless the sacrifice is spread around fairly, like rationing in World War II. People look at Shell and Exxon and BP and don't see fairness."

"It's working people who are losing jobs now in every industry involved with fossil fuels, and that's—"

"Don't lecture me, Kayla. I know the statistics." He looked ten years older than he had in January.

When had her disdain for him turned to admiration? She didn't know.

The country plunged into a recession. Stocks fell. Unemployment rose. Miscarriages rose. Political campaigns turned dirtier than anyone living had ever seen, with attempts to block registration and highly illegal door-to-door intimidation of voters. Someone shot and killed a green candidate in Louisiana, where oil drilling was a major industry. The shooter then killed herself, gaining a sort of martyrdom.

Lawsuits and countersuits clogged court dockets: restraint of trade, wrongful death, torts as varied as microbes. Rich people fled the PV-45-infected states, and the mosquitoes fled north right along with them. Except in the states on the Canadian border, autumn weather did not kill *Aedes*. Global warming had made its extended range possible.

LEAVE THE OIL IN THE SOIL.

SAVE THE CHILDREN!

KILL PV-45 TERRORISTS, NOT OUR JOBS.

NO GREEN FASCISM!

FUCK SHELL—IT FUCKED US.

NO MORE GREEN LIES!

Some days it seemed to Kayla that the only thing the country was producing was slogans.

Then a protest at a drilling site in Colorado turned violent, and the governor sent in the state militia. Seventeen protesters and three soldiers died. Kayla found James with his head in his hands. Stunned, she put her arms around him. He clung to her like a child but said nothing. After a moment, he rose, went into another room, and closed the door.

***

Kayla and James got out of the car and walked across the tarmac to his corporate plane, which he had finally begun to use regularly. They were leaving Dallas's Love Field for a meeting in DC. Very hush-hush, very important. The election was three days away.

Planes bellied up to jetways. One taxied by on the closest runway. Baggage carts drove through the heat, stifling even in November. Kayla smelled jet fuel in the air.

Toluene. Half of 1 percent of jet fuel, 15 percent of gasoline. Her head was a cache pond of percentages. 75 percent of chemicals rated by the EPA had never been tested on fetuses or children, 11 percent of people live within one hundred meters of a major highway, 25 percent of Germany's energy comes from renewables, investments in public transit create 31 percent more jobs than in new roads and bridges... *Stop*. She hadn't slept in twenty-six hours. She had no idea when James had last slept.

The baggage handler leaped from his cart and sprinted toward James. Kayla saw him clearly. Young, scowling, goggled, and earplugged. He aimed a nine-millimeter at James and fired.

James went down. Another shot. Kayla hit the ground. The assassin was already dead, and James's bodyguards swept the area, looking for more shooters. There weren't any.

Kayla bent over James. "James?"

"It's…nothing…"

It wasn't nothing. Sirens screamed. In the ambulance, James fainted. Kayla waited outside the OR, holding his briefcase. She studied the pattern on the floor tiles, squares and circles in a bilious shade of green.

*Let him be all right.*

Squares.

*He's so close.*

Circles.

*And so am I.*

She had a new lead, one she was not going to tell him about, not now. Maybe not ever, depending on what she found.

"Ms. Serdenovna? Are you a relative of Mr. Sullivan's?"

"Yes," she lied.

"He'll be fine. He's in recovery now, and you can see him soon."

As soon as he was out of recovery, James signed himself out of the hospital against medical advice. He refused to talk to the press. Kayla stopped reading or listening to the resulting avalanche of reporting, most of it wrong. The next day, bandaged and on painkillers, he took his meeting in DC.

"James, after the election, I need a few days away."

"*Now?*"

"Yes. Now. It's personal."

He blinked, then nodded. "Certainly. Where do you want the plane to take you?"

"I'll make my own arrangements."

He looked impatient. "That's silly. Let me help. And you'll be back here faster. Things are happening."

"I *said* that I'll make my own arrangements."

He looked at her hard. "Have it your way."

***

At the polls, it went James's way.

Kayla felt a strange sense of unreality as she sat with him and his senior staff in a hotel room with a TV the size of a Tudor armoire. He had done it. At the cost of untold billions—although he would tell her, she knew, the exact number once he had it—James Conway Sullivan had bought himself a Congress that would wrench the entire economy in a new direction. Green Republicans—there were some—green Democrats, green Greens. A chartreuse coalition for a country in recessed turmoil, a coalition charged with pulling it out of tar pits and into windy, watery, sunny renewables, and with doing that fast enough to avoid sinking everything. Not, however, without bitter fights. Everyone knew that. But with disease-panicked support from below, James had changed the lawmakers so that they could change the laws.

Kayla congratulated James, went to her room, and slept for eleven hours. Two days later, she boarded an American Airlines flight for Minneapolis. The flight was less than one-third full, and waves of demonstrators on both sides of the green divide surged around and nearly into the airport. But the planes were flying.

***

She was almost too late, on an errand that had no real point except to mask her second plane trip, from a different airport in a different city, from James. Or maybe, like his famous software, she needed three-dimensional reality instead of just an image in her mind.

The baby still lay in the nursery only because Jenna had had to deliver by C-section. Mother and baby, said a chatty nurse to Jenna's "cousin," were going home tomorrow. Kayla did not see Robert. She gazed through the wide glass window at the rows of bassinets holding newborns asleep, or crying, or gazing at that marvelous creation, the ceiling. "Baby Girl Patterson" was a gazer. Wide blue eyes, wispy hair the color of Robert's, a pink blanket. She flailed one impossibly tiny fist.

*Being a mother is all my Mary ever wanted. Her heart's broken, and mine, too.*

So many Marys. So many kinds of heartbreak.

She left the hospital and caught her flight to Miami.

***

In January the new radical Congress convened, facing a country in economic free fall. By the end of February, the Big Three auto companies had received bailouts, but with strings as heavy as cables. They were required to install the new carbon-sucking machines on their factories and to invest massively in electric cars and biofuels. The new mantra in DC was "the polluter pays."

With cars more expensive, mass transit was built faster than anyone thought possible, funded by a middling carbon tax of fifteen dollars per metric ton. Industry wanted no tax; the congressional majority wanted forty. The compromise was followed by screaming and yelling on both sides, but the public mood was definite: America was moderating its love affair with autos. *Save the Children.*

Fracking bans went nationwide; even before PV-45, studies proved that near major fracking sites, the miscarriage rate had been double the national average. Now the rate was over 90 percent.

Money to switch energy sources to renewables came from multiple sources: low-rate transaction taxes on the trading of stocks and derivatives. Higher royalty rates on oil, gas, and coal extraction. The wealthiest 5 percent of the country faced much steeper income taxes, and many loopholes were closed. Luxury taxes funneled money to cities and towns, earmarked exclusively for renewables. Those states whose budgets depended heavily on severance taxes from fossil-fuel extraction were compensated by a complicated formula that almost no one understood but that seemed to work, although not at first. More crises in Colorado, North Dakota, Wyoming. Jobs were lost, but as work on new infrastructure began on a massive scale, people again found employment.

Key to making it all succeed, as James had insisted to Kayla from the beginning, was limiting fossil fuel's ability to elect and control politicians. "Politicians don't say no to powerful corporations. At best, they say 'Maybe' or 'I'd like to' or 'We need a good study on that.'" Now political donations were tightly capped and full disclosure required, with heavy penalties for violations.

The lawsuits went international. The World Trade Organization filed several, alleging restraint of trade. It *was* restraint of trade; corporations could no longer move their factories overseas, creating pollution and cheap-labor sweatshops in other parts of the world, without facing ruinous penalties. The WTO also said that member countries could not privilege domestic trade over foreign trade, which meant no subsidies to companies producing renewable energy products. No local subsidies (the WTO had been suing over this for decades), no state subsidies, no national subsidies.

The United States did the unthinkable: it withdrew from the World Trade Organization.

"Protectionism never works," Kayla said to James.

"Not long term, no. We need it short term, for this transition in the United States. The others will follow. Congress is investing in third-world development through renewables. Give it time, Kayla."

"How much time?"

"Do I look like Chronos? I don't know. God, you're impatient. How long did it take to pull the world out of the Great Depression?"

"It took a war."

He was silent. It was what everyone feared: that America's pushing so hard would lead to war. Kayla, everything inside her roiling, turned toward the door.

James said, "Why are you spending so much time in Florida?"

"Are you having me watched?"

He didn't answer that. "How is your book coming?"

"I'm waiting for the ending." She turned her head, one hand on the doorknob, to give him a hard, level stare. But all he said was, "There is no ending. It's a process."

"Right. Uh-huh. I'll wait anyway."

***

There was no war. By the presidential election, two years later, there was a vaccine for PV-45. The country elected a green president anyway and returned most of the current Congress. Wind farms and solar panels, both with much higher efficiency resulting from heavy investment, produced half the country's electricity, with the percentage increasing steadily. People in sunny or windy areas sold power back to the grid, a wildly popular source of supplementary income. Unemployment was low. Taxes were high, but incomes were rising. Necessary traffic flowed easily on radically less–congested highways. Many of those vehicles, more every year, ran on either electricity or biofuels made without toluene. Mag-lev trains crisscrossed the country. In the cities, fast and reliable mass transit was still being built incredibly quickly by old standards.

There was a long way to go, especially in states whose economies had leaned heavily on fossil fuels. There were pockets of protest, of defiance, legal battles. The president had won election by only a narrow margin.

Other countries, forced by the example and political pressure and outright economic bribes from the United States, were grudgingly and slowly following suit. Global warming slowed as emissions fell. China, particularly, accelerated its program to clean its barely breathable air and sludgy rivers. "It's always easier," James said dryly, "to get programs in place when you don't need to pay attention to anyone's civil rights."

Kayla's contract with James was finished, and by March her book was completed except for the last chapter. Before she could write it, she needed to talk to James.

Maybe, she hoped desperately, she was wrong.

She flew to Flint.

***

James had bought a modest, heavily guarded house in the countryside. To some, he was a hero; to others, Satan incarnate. He gave no interviews, made no public appearances. Internet rumors said he was dead, or gone insane, or taken by aliens "back home."

Kayla navigated a series of checkpoints and electronic surveillance to get into the house. James met her in the library, a small room lined with freestanding bookcases, four chairs in the middle around a low table. Kayla had seen the same chairs on sale at Macy's.

"Hello, Kayla."

She was shocked at his appearance. Always thin, now he looked skeletal, his sunken eyes shadowed. She blurted, "James, did you lie to me about being sick?"

"I never lied to you."

Her anger flashed. "But there's a lot you withheld from me!"

He ignored this. "Are you here to bring me the last part of the book?"

"No. To write it."

If her answer surprised him, he didn't show it. Kayla handed him a thin sheaf of papers; this wasn't something she wanted in digital, and therefore pirate-able, form. He read in silence.

"How did you get this?"

"It doesn't matter how I got it. I'm a journalist, remember?"

"And I gave you a virtually unlimited expense account. Plus enough access to leads that you figured out where to look. Which is why you went to the Caymans more than once." It wasn't a question.

She found she was trembling and clenched her ass cheeks, an old trick. She was not going to tell him anything, just as the CDC had not revealed what they had at least suspected. "You bought a struggling, offshore biotech lab whose products never could have met with FDA approval. You bought it from some pretty shady sellers, in total contrast to your usual squeaky-clean business deals. When something doesn't make financial sense, I've found, it usually makes some other kind of sense.

"James...you created PV-45. In that lab in the Caribbean. You had the virus genetically engineered to cause liver changes...to cause all those miscarriages. And when PV-45 was released in Louisiana, in that upscale suburb, you only pretended to think it was a setback to your plans."

"Yes."

"Just to—"

All at once he was across the room and clutching her arm with more fierce strength than she'd thought he could possess. She had never seen him angry before.

"Yes! 'Just' to remake the economy. 'Just' to abandon the fossil fuels that built this country before those same fuels had a chance to destroy it. 'Just' because our collective survival as a species depends on stopping the climate changes that might wipe us out. 'Just' because even if it doesn't get that bad, millions if not billions will die from flooding, food shortages, and all the rest of it while a tiny fraction of humanity fiddles around with compost heaps and roof gardens and thinks it's accomplishing something…'just' that!"

Kayla wrenched herself free. "And what about the risk that we wouldn't do your 'remaking' and wouldn't find a vaccine, so that your disease destroyed the next generation?"

"Wasn't going to happen. 'Save the Children.'"

"All those lost jobs, all those families struggling to have kids and bitterly disappointed…what about compassion?"

"This is compassion. Anyway, you're asking the wrong question." He walked away from her and picked up a glass from the table, its rim sticky with whiskey.

"What's the right question?"

"The right question is why I had to do it this way. Why I took it on myself to play God."

"All right—why did you?"

"Because nothing else would have worked."

Silence. He downed his drink in a single long swallow and swiped his hand across his mouth in a crude, uncharacteristic gesture. His next words were quieter.

"People want change, but they don't want to pay the real costs, and major change *always* costs. And I didn't have the luxury of time. In another fifteen years, maybe ten, it would have been too late. You've seen the warming curves. You know when the feedback loops kick in and climate change can't be stopped."

She did. She said, "I'm going to write about this," and for a sudden sickening moment she remembered the tarmac bodyguard with deadly aim. *Am I in danger?*

No. This was James. She knew him, even if not as completely as she had vainly assumed. *Dragons,* she thought—there were all kinds of dragons, including those of the mind: fierce, primitive, sometimes untamable thoughts and desires. Even if a story contained no physical dragons, dragons were there. They were always there.

James said, "So this is the last chapter of your book. I always knew that someone would discover the lab. I'm glad it was you."

"James—they'll crucify you."

"Interesting choice of words." He smiled, an expression so complex that Kayla would remember it for the rest of her life: pain, humor, resignation, regret.

She blurted, "You're a monster. And the bravest man I know."

He said with sudden ferocity, "It *was* the only thing that would work!" In a moment, the fierceness had vanished. He set down his glass. "Excuse me a minute—bathroom. Too much whiskey." In the doorway, he said over his shoulder, "The body, unfortunately, does have more limits than the mind."

He closed the bathroom door. Kayla heard it lock. A moment later, she heard the single shot.

*Thanks to the Year Without a Winter conference, which not only provided information for this story but also fresh ways to think about the problem of global warming.*

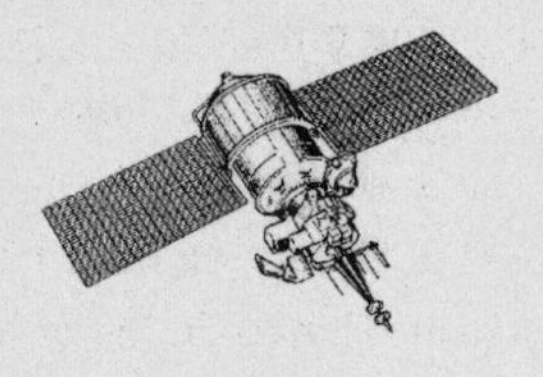

# Mother of Invention

*Nnedi Okorafor*

"Error, fear, and suffering are the mothers of invention."
—Ursula K. Le Guin, *Changing Planes*

It was a beautiful sunny day, and yet Anwuli knew the weather was coming for her.

She paused on the lush grass in front of the house, purposely stepping on one of the grass's flowers. When she raised her foot, the sturdy thing sprung right back into place, letting out a puff of pollen like a small laugh. Anwuli gnashed her teeth, clutching the metal planks she carried and staring up the driveway.

Up the road, a man was huffing and puffing and sweating. He wore a clearly drenched jogging suit and white running shoes that probably wanted to melt in the Nigerian midday heat. Her neighbor Festus Nnaemeka. The moment she and Festus made eye contact, he began walking faster.

Anwuli squeezed her face with irritation and loudly sucked her teeth, hoping he would hear. "Don't need help from any of you two-faced people, anyway," she muttered to herself, watching him go. "You keep walking and wheezing. Idiot." She heaved the metal planks up a bit, carried them to the doorstep and dumped them there. "Obi 3, come and get all this," she said. Breathing heavily, she wiped sweat from her brow, rubbing the Braxton Hicks pain in her lower belly. "Whoo!"

One of Obi 3's sleek blue metal drones zipped in and used its extending arms to scoop up the planks. The blown air from its propellers felt good on Anwuli's face, and she sighed.

"Thank you, Anwuli," Obi 3 said through the drone's speakers.

Anwuli nodded, watching the drone zoom off with the planks to the other side of Obi 3. Who knew what Obi 3 needed them for; it was always requesting something. Obi 3 was one of her now ex-fiancé's personally designed shape-shifting smart homes. He'd built one for himself, one for his company, and this third one was also his, but Anwuli lived in it. And this house, which he'd named Obi 3 (not because of the classic *Star Wars* film but because *obi* meant "home" in Igbo, and it was the third one), was his smallest, most complex design.

Built atop drained swamplands, Obi 3 rested on three mechanized cushioning beams that could lift the house up high when it wanted a nice view of the city, or keep it close to the ground. The house could also rotate to follow the sun and transform its shape from an equilateral triangle into a square and split into four separate modules based on a mathematical formula. And because it was a smart home, it was always repairing and sometimes building on itself.

Over the past five months, Obi 3 had requested nails, vents, sheet metal, planks of wood, piping. Once it even requested large steel ball bearings. Paid for using her ex's credit card, most of the time she just had it delivered and dropped at the doorway, or she'd pick up the stuff and place it there, where she quickly forgot about it. By the time she came back outside, it was always gone, taken by the drones. None of this mattered to her, though, because she had real problems to worry about. Especially in the last eight months. Especially in the next hour.

"Shit," she whimpered, holding her very pregnant belly as she looked at the clear blue sky, again. There had been no storms in the damned forecast for the next two weeks, and she thought she *had* finally been blessed with some luck after so long. However, apparently the weather forecast was wrong. Very, very wrong. She felt the air pressure dropping like a cold shiver running up her spine. Mere hours ago, Dr. Iwuchukwu had informed her that this sensitivity to air pressure was part of the allergy.

Several honeybees buzzed around one of the flowerbeds beside her. The lilies and chrysanthemums were far more delicate than the government-enforced supergrass, but at least they were of her choosing. Just as it was her choice to stay in *her* house. She listened harder, straining to hear over the remote sound of cars passing on the main road a half-mile away. "Dammit," she whispered, when she heard the rumble of thunder in the distance. She turned and headed to the house.

The door opened, and she went inside and slammed it behind her before it could close itself. She stood there for a moment, her hands shaking, tears tumbling down her face. The house had drawn itself into its most compact and secure shape, a square, swinging the triangular sections of the kitchen and living room together. Outside, from down the road, the mosque announced the call to prayer.

"Fuck!" she screamed, smacking a fist to the wall. "*Tufiakwa*! No, no, no, this is *not* fair!" Then the Braxton Hicks in her belly clenched, and she gasped with pain. She went to her living room, threw her purse on the couch, and plopped down next to it, massaging her sides.

"Relax, oh, relax, Anwuli. Breathe," Obi 3 crooned in its rich voice. "You are fine; your baby is fine; everything is fiiiiiine."

Anwuli closed her eyes and listened to her house sing for a bit, and soon she calmed and felt better. "Music is all we've got," she sang back to Obi 3. And the sound of her own voice pushed away the fact that she and her baby would probably be dead by morning, and it would be all her fault. Pushed it away some.

Music and Obi 3. Those were all she and her unborn baby had had for nine months. Since she'd learned she was pregnant and stupidly told her fiancé, who a minute later blurted to her that he was married with two children and couldn't be a father to her child, too.

The city of New Delta was big, but her neighborhood had always been "small" in many ways. One of those ways was how people stamped the scarlet badge of "home-wrecking lady" on women who had children with married men. Her fake fiancé had deserted her, using the excuse of Anwuli playing the seductress he couldn't resist. Then her friends stopped talking to her. Even her sister and cousins who lived mere miles away blocked her on all social networks. When she went to the local supermarket, not one person would meet her eye.

Only her smart home spoke (and sometimes sang) to her. And then there was the baby. Boy, girl, she refused to find out. It was the only good thing she had to look forward to. But her baby was making her sick too, specifically allergic. Dr. Iwuchukwu had been telling her to leave New Delta for months, but Anwuli wasn't about to leave her house. The house was her respect; what else could she claim she'd earned from the relationship? She knew it was irrational and maybe even deadly, but she took her chances. So far, so good. Until today's diagnosis at her doctor's appointment. And right there in that antiseptic place, whose smell made her queasy, she'd decided for good: she wasn't going *anywhere*. Come what may. Now, as if the cruel gods were answering her, a storm was coming.

"Seriously," she muttered, sinking down on the couch, letting its massagers knead the tight muscles of her neck. "I have such bad luck."

"Bad luck is only a lack of information," Obi 3 said. "Dr. Iwuchukwu has sent you a message saying to go over it again."

"I understood it the first time," she said. "I just don't care. I'm not going anywhere. The idiot left me. He's not getting his house back, too."

Before Anwuli could launch into a full-blown rant, Obi 3 began playing the informative video the doctor suggested. She sighed with irritation as the image opened up before her. She didn't care to know more than the bits her doctor had told her, but she was tired, so she watched anyway.

The man walked with a cane and wore an Igbo white-and-red chief's cap like an elder from Anwuli's village in Arochukwu. The projection made it look as if he walked in from the bedroom door, and Anwuli rolled her eyes. This entrance was supposed to be more "personable," but she only found it obnoxious.

"Hello, Anwuli," the man said, graciously. "So, you live in New

Delta, Nigeria, the greenest place in the world. Fun fact: a hundred years ago, this used to be swamplands and riverways, and the greatest export was oil. Violent clashes between oil corporations and a number of the Niger Delta's minority ethnic groups who felt they were being exploited..."

"Skip," Anwuli said. The man froze for a moment and went from standing in the living room to standing in the middle of downtown New Delta. Anwuli was about to skip again, but instead she laughed and watched.

In the area between New Delta's low skyscrapers, buildings and homes were carpeted with its world-famous stunning green grass, and the roads were fringed with it, but in this scene the grass was covered with smiley-faced bopping periwinkle flowers. It looked ridiculous, like one of those ancient animations from the early 1900s or a psychedelic drug-induced hallucination. The man grinned as he grandiosely swept his arms out to indicate all the lush greenery around him.

"Grass!" he announced. "Whether we know it or not, grass is important to most of us. Grass is a monumental food source worldwide. Corn, millet, oats, sugar—all of them come from grass plants. Even rice was a grass plant. We use grass plants to make bread, liquor, plastic, and so much more! Livestock animals feed mostly on grasses, too. Sometimes we use grass plants like bamboo for construction. Grass helps curb erosion."

He walked closer and stood in the center of town square in the grassy roundabout, smart cars and electric scooters driving round him. At his back stood the statue of Nigeria's president standing beside a giant peri flower. "The post-oil city New Delta is now the greenest place in the world, thanks to the innovative air-scrubbing superplant known as periwinkle grass, a GMO grass created in Chinese labs by Nigerian scientist Nneka Mgbaramuko.

"Carpeting New Delta, periwinkle's signature tough flowers are a thing of beauty and innovation. A genetic hybrid drawn from a variety of plants including sunflowers, zoysia grass, rice, and jasmine flowers, we can thank periwinkle grass for giving us the perfect replacement for rice just after its extinction. The grass produces periwinkle seed, more commonly just called 'peri,' which is delicious, easy to cook, quick to grow. And it can grow only here in New Delta, because of the special mineral makeup from its past as a swamp. What a resource!" He held up a hand, and the point of view zoomed in to the soft light-purple-blue flower in it. The man looked down at Anwuli as he grinned somewhat insanely. "One week a year, the harvester trucks come out to—"

"Ugh, skip," she said, waving a hand. "Just go to 'New Delta Allergies.'"

The man froze and then reappeared in what looked like someone's nasal cavity, the world around him red and smooth.

"Allergies," he said, looking right at Anwuli with a smirk. He winked mischievously. "Humans have had them since humans were humans, and maybe before that. One of the earliest recorded incidents was sometime between 3640 and 3300 BC when King Menses of Egypt died from a wasp sting.

"In New Delta, pollen allergies are commonplace. Milder symptoms include skin rashes, hives, runny nose, itchy eyes, nausea, and stomach cramps. Severe symptoms are more extreme. Swelling caused by the allergic reaction can spread to the throat and lungs, causing allergenic asthma or a serious condition known as anaphylaxis.

"New Delta is a wonderful place of spotless greenery where one can walk about with no shoes on the soft grass, breathe air so clear it smells perfumed, and drive down Nigeria's cleanest streets."

At this Anwuli laughed.

"But in the last five years, due to an unexpected shift in the climate, pollination season has become quite an event. This means more copious harvests of peri. But because peri grass is a wind pollinator, it also means what scientists have called 'pollen tsunamis.'" The weather around the man grew dark as storm clouds moved in and the room vibrated with the sound of thunder. Anwuli glanced toward the side of the room that was all window. Outside was still sunny, but it wouldn't be for long.

"Skip to Izeuzere," she said.

The man froze and then was sitting behind a doctor's desk, wearing a lab coat. He still wore his Igbo chief cap. "...a few New Delta citizens were diagnosed with an allergy called Izeuzere. The name, which means 'sneeze' in Igbo, was given to the condition by a non-English-speaking Igbo virologist who liked to keep things simple.

If someone with Izeuzere is caught in a pollen tsunami, there will first be severe runny eyes, sneezing fits, and then an escalation to convulsions, 'rapid rash,' and then suffocation. Most who have it experience a preliminary sneezing fit and then the full spectrum of symptoms the moment a pollen tsunami saturates the area. Deadly exposure to the pollen when a tsunami hits takes minutes, even when indoors, and is instant when outside. Treatment is to leave New Delta and go to an arid environment before the next pollen tsunami. Once there, one must be given a battery of anti-allergen injections for five months."

"What if I lose everything if I *leave*?" Anwuli asked the virtual man. "What if moving out of this house allows the father of my child to get rid of me without lifting a damn finger? Do you have answers for that in your database?" The man's eyebrows went up, but before the man could respond, she screamed, "Shut up!" She punched the couch cushion. "Off! Turn off!" The image disappeared, replaced by her favorite soothing scene of an American cottage covered in snow. The sound of the wind was muffled by the blanket of snow, and smoke was rising from the cottage's chimney. She knew what would happen if she couldn't leave the area. "Dammit," she hissed. "I refuse! I *refuse*!"

"Are you sure you don't want me to buy a ticket for you to Abuja?" Obi 3 asked. "There is a flight leaving in two hours. Your auntie will..."

"No!" Anwuli sat back and shut her eyes, feeling her frustrated tears roll down both sides of her face. "I'm not leaving. I don't care." She paused. "They probably all hope I'll die. Like I deserve it."

"What kind of dessert would you like? There is caramel crème and honeyed peri bread."

"*Deserve,*" Anwuli snapped. "I said *deserve*, not *dessert!*"

"You deserve happiness, Anwuli."

Anwuli closed her eyes and sighed, muttering, "Left me alone here for nine months; their message is clear. Well, so is mine. I'm not going. This is his baby. He can't deny that forever." She paused. "Now this stupid storm rolls in out of nowhere when I could have this child at any moment. This is God's work. Maybe he wants all my trouble over with fast."

"Would you like some jollof peri and stew?" Obi 3 asked as Anwuli slowly got up. "You haven't eaten since before you went to your appointment."

"Why would I want to eat when I am about to die alone?" she shouted.

She got up. She stared around Obi 3. Not spotless, because Anwuli didn't like spotless, but tidy. *Her* space since *he* had left her to fully return to his marital home. One of Obi 3's interior drones zipped into the bedroom with a set of freshly washed and folded clothes.

"What do I do?" Anwuli whispered. And, as if to answer, the sound of thunder rumbled from outside, this time louder. "I don't want to die."

She'd always had allergies. Her father had even playfully nicknamed her "*ogbanje*" when she was little because she was always the one sniffling, sneezy, and sent to bed anytime the peri flowers bloomed. Goodness knew that when her allergies flared, she did feel like a spirit who'd prefer to die and return to her spirit friends than keep living with the discomfort. But *never* did she imagine she'd eventually come down with the rare illness everyone had been talking about. And her doctor, also a local to her community, had been so cold about it.

"I don't know why you haven't left yet, but don't worry. You'll give birth any day now," he'd said at her earlier appointment, clearly avoiding her eyes by looking at his tablet. "Then you take your baby, fly to Abuja immediately and get treatment there. No storms are due in the next week, so you will be fine."

Anwuli had nodded agreement. What she didn't say in that room was that she had no intention of leaving. Obi 3 was her home as long as she lived in it. Bayo was an asshole, but he could never throw her out of the house, no matter how much he wanted the situation to go away. She was sure he still loved her, and above everything, this *was* his baby. However, his wife certainly would love for her and his "bastard" baby to simply leave the area. But none of it mattered now because here was the thunder.

Anwuli went to her room and curled up in her bed and for several minutes, minutes she knew would be her last, she cried and cried. For herself, for her situation, her choice, for everything. When she couldn't cry anymore, the thunder was closer. She got up. Her belly felt hard as a rock, and the pain drove even her fear of death away. At the same time, Obi 3 brightened the lights, which seemed to amplify the pain.

"Blood of Jesus!" she screamed, crumbling to the floor in front of the couch.

She was twenty-nine years old and she'd watched all her friends settle into marriage and have child after child, yet this was her first. And there had been so much chaos around the fact of her pregnancy that although she went to regular checkups, she hadn't really thought much about the birth or what she'd do afterward. Shame, desperation, embarrassment, and abandonment burned hotter and shined brighter than her future. So Anwuli wasn't ready.

Now her pain had begun to speak, and it told vibrant stories of flesh-consuming fire that burned the body to hard hot stone. It was as if her midsection was trying to squeeze itself bloody. She rolled on the floor, more tears tumbling from her eyes. And then…it passed. Her belly melted from hot stone back to flesh, her mind cleared, and a light patter of rain began tapping at the windows.

"Better?" Obi 3 asked.

"Yes," Anwuli said, grasping the side of the couch to pull herself up. Beside her hovered one of Obi 3's drones. "I'm OK. I can do it myself."

"That was a contraction," Obi 3 said. "The variations in electromagnetic noise my sensory lights are picking up tell me that you'll be entering labor soon."

Anwuli groaned, glancing at the window. *Of course*, she thought.

"Not yet but very soon," Obi 3 said. It beeped softly, and the lights flashed a gentle pink orange. "You have a phone call. Bayo."

Anwuli frowned. She shut her eyes and took a deep breath. "OK, answer."

There was another beep, and Bayo's face appeared before her. He looked sweaty, and his shaven brown head shined in the light of the room he was in. He squinted. "Anwuli, turn your visuals on," he said.

"No," she snapped, propping herself against the couch. "What do you want?"

He sighed. "Your doctor just called me."

"What did he say?" she asked, gnashing her teeth.

"That you're sick. That you have…Izeuzere. How can this be? Is it the pregnancy?"

"Is it legal for him to discuss confidential patient information with strangers?" she snapped. "Doubtful."

"I'm not a stranger."

"The last time you spoke to me was nine months ago."

His shifty eyes shifted. There was a shadow beside him; someone else was there. Probably his wife. Anwuli felt a wave of wooziness pass over her. "I…I think I'm in labor," she said.

He looked surprised but then shocked her by saying nothing.

"No ambulance will drive through this storm," she said. "Can you…?"

His wife's face suddenly filled the virtual screen. "No, he will *not*," she said. "He has a family and cannot afford to go driving into pollen storms. Clean up your own mess. And get out of *our* house!" Bayo's wife continued to block Bayo's face, and if Bayo said anything, Anwuli could not hear him.

"Whose house?" Anwuli shouted at her. "Did *you* design it? Build it? Pay for it? Does this house even know your name?"

"Go and *die!*" his wife roared. The image disappeared.

Anwuli flared her nostrils, but no effort could stop the tears and hurt from washing over her like its own contraction. She hadn't known a thing about that woman. Bayo had. Yet who did his family and the rest of the community embrace? Who still had his own body to himself? *Well*, Anwuli thought, *maybe I* did *know about her. Maybe. Let me not lie when I am so close to my death. I knew. I just chose not to see.*

"Call parents," she breathed.

Their phone rang and rang. No response. Not surprising. They'd stopped picking up her calls months ago. She sent a text explaining it all, then went to the kitchen.

***

Sitting at her large dinner table, she'd just finished a plate of jollof peri (delicious, even if the thousands of acres of it outside were going to kill her), roasted goat meat and boiled yam and palm oil, when the next contraction smashed into her. The lights brightened again, and a drone was in front of her with a bowl a second before she threw up everything in her stomach.

The strain of throwing up *and* having a contraction nearly caused her to pass out. One of Obi 3's drones pushed itself beside her to keep her from tumbling to the floor.

"The variations in electromagnetic noise my sensory lights pick up alert me that..."

"Shut up!" Anwuli screamed.

"...you are now in labor," Obi 3 finished.

*Boom*! the thunder outside responded. Sheets of rain began to pelt Obi 3. The lights flickered, and then Anwuli heard her backup solar generator kick in.

"What do I do?" she grunted, using a napkin to wipe her mouth. "What am I going to do?"

"Shuffling songs by MC Do Dat," Obi 3 cheerily said. Bass-heavy rap music shook the entire house, making Anwuli even more nauseous.

"Ladies do dat,

Bitches do dat!

Get down low and,

Do dat, do dat!" MC Do Dat rapped over the beats in his low raspy voice.

"Music off!" she screamed, tears squeezing from her eyes. She clenched her fists with rage. "No music! Ooh, I *hate* this song!"

The music stopped in time for the sound of thunder to shake the house. Anwuli slowly dragged herself up as the contraction subsided.

"I'll help you to the couch," Obi 3 said.

She nodded and leaned against the drone that floated to her side. As she did, reality descended on Anwuli. Obi 3 was only an extension of herself. She was only talking to herself, being helped by herself. She was alone. "The storm...pollen...I don't want to..." As she stumbled to the couch, the drone holding her under her armpit, she started to cry again. She cried more as she fell onto the couch and rolled onto her back, her clothes now drenched in sweat. She cried as she stared at the spotless sky-blue ceiling, which she had used Obi 3's drones to paint when Bayo left her. She cried as lightning flashed and the thunder roared outside, the unpredicted storm's winds blowing.

She'd been crying for nine months, and she cried for yet another ten minutes, and then another contraction hit, and she forgot everything. As the minutes passed, and the contractions came faster and faster, she didn't remember where the pillow came from that propped her up or how her legs held themselves apart. What she did recall was the window across from her shattering as a palm tree fell through it. She remembered the wind and rain blowing into Obi 3, filling it with the heat and humidity from outside. Tree leaves, new and dry, slapped against the couch, onto the floor, but no peri flowers were blown in. Those were strong like men; they didn't even lose petals in the worst wind. *Built to survive and reproduce, not to keep from killing us*, she vaguely thought. She couldn't help but note the irony: plant fertilizer was going to kill her as she was giving birth.

Her face grew damp with sweat and rain. As she gave the great push that thrust her first child into the world, the storm outside exhausted itself to a hard rain. The coming of her child felt like her body submitting after a battle. The sharp pain peaked and then retreated. And that is how the first to carry her squirming daughter was not a human being but a drone, using a plastic scooper as its long sharp knife cut the cord. When the drone placed the child in Anwuli's hands, she looked down at her daughter's squashed, agitated face. For several moments, she stared, unmoving.

"Don't you want to cry?" she asked the snuffling infant.

"Mmmyah," the baby said, turning her head this way and that. Anwuli found herself smiling. She poked her daughter's little cheek. The moment she felt the baby's softness, Anwuli began to weep. She touched the baby again, running a finger delicately across the baby's cheek to touch her lips. Immediately, her child began to suck on her finger.

"She's breathing strongly already," Obi 3 said. "Maybe she does not need to cry."

"Mmiri," Anwuli said, holding the child to her. "I'll name you Mmiri. What do you think, Obi 3?"

"Mmiri means 'water' in the Igbo language," Obi 3 said.

Anwuli laughed. "OK. But do *you* approve?"

"You do not need my approval to name your child."

"But I would like it, if you think to give it."

There was a pause. Then Obi 3 said, "How about giving her the middle name Storm? Storm was the American Kenyan superheroine from Marvel comics. She could control the weather and fly."

Anwuli's eyebrows rose. "Hmm, wow," she said. "Mmiri Storm Okwuokenye, then. I approve." The house glowed a soft lavender color that turned the ceiling a deeper sky blue as Anwuli stared up at it. "Mmiri Storm Okwuokenye," Anwuli breathed again, looking at her new daughter, who smelled like the earth. Bloody, coppery, yeasty. Hers. She held on to this beautiful thought and the sound of her daughter snuffling as the pains of expelling the afterbirth came. When this was over, she slumped on the couch, watching the drone take away the bloody mass.

She already felt much better. Then she sneezed, and her eyes grew itchy. "No," she whispered. Baby Mmiri decided it was time to start wailing. The rain had stopped, and the sun was already peeking through the retreating clouds. She sneezed again, and the house drone flew to her, a clean orange towel now draped over its scooper. Anwuli put her daughter into it and was wracked by a sneeze again. She sat up, surprised by how OK she felt. The second drone flew up beside her carrying a glass of water and her bottle of antihistamine tablets. "Hurry, take three," Obi 3 said. "Maybe—"

"There is nothing to be helped now," Anwuli blurted, looking at the shattered window. Already, what looked like smoke was wafting into the house. Soon visibility outside would be zero, and it would last for the next twenty-four hours. "I'm a dead woman."

No one had predicted weather patterns shifting. This is why scientists were calling the occasional spontaneous variation in weather patterns "climate chaos" instead of "climate change." That's what they'd recently been saying on the news, anyway. The pollinating grass was genetically staggered to release pollen at three separate times during the year, with one-third of the grass pollinating in each period. However, over the last twenty years, an unexpected shift in the length of the dry, cool Harmattan season had scrambled that timing, causing the pollination periods of all three groups to align.

The immense wealth made from peri production went directly to the Nigerian government and to the Chinese corporations who'd invested so deeply in Nigeria for decades, and next to nothing went to

New Delta, much in the same way it had when the greatest resource had been oil. For this reason, the initially lovely city that was New Delta began to deteriorate, and the Chinese and Nigerian governments paid less attention to the pollination misalignment. News of pollen allergies had become nationally known only when Izeuzere set in during the last two years. But only because the way it killed was so spectacular. And this year, rainy season had been particularly wet.

"I'm dead," Anwuli muttered, using all the effort she could muster to get up. She threw her legs off the couch, planting them on the floor. Ignoring the blood soaking her bottom through the drenched towels, she pressed her fists to the cushions on both sides of her. Then she lifted herself up. The pain was far less than she expected, and she froze for a moment, glad to be on her feet.

"Standing," she whispered, her nose now completely stuffed and her eyes still watering. She sniffed wetly. Her insides felt as if they would plop out between her legs onto the blood-spattered carpet. But they didn't. She touched her deflated belly. Then she sneezed so hard that she sat back down. In the kitchen, her baby was crying as the drone put her in a tub of water to wash her off. Anwuli pushed herself up again and took a step toward the kitchen. But as she took another, her chest grew stiff. She wheezed.

She couldn't tell if the room was blurry because it was full of pollen or because of her watering eyes or the fact that she could barely take in enough oxygen. And then she was falling. As she lay on the floor, she heard Obi 3 talking to her, but she didn't understand. Her baby was crying, and if she could smile, she would have because her baby was not sneezing. Then she closed her eyes, and it was as if the world around her was breaking.

***

The floor shook, and Anwuli heard the walls cracking, shifting, crumbling. Her nose was too stuffed for her to smell anything, but she could feel pollen coating her tongue and blood seeping from between her legs. Things went black for a while. Mmiri's cries faded away and stopped. The noise of things breaking became a low hum. The shaking stopped.

Anwuli must have slept.

She sneezed hard and wheezed, cracking her gummy eyes open. Everything was a blur until she blinked. She gasped. Then she realized that she *could* gasp. And the room was suddenly warm, like outside. She blinked several more times, wiped her eyes, and then just stared at where the broken window had been. Her daughter began weakly crying. The makeshift cradle the drone had begun to gently rock, and Mmiri quieted a little.

Still staring but slowly sitting up, Anwuli said, "Bring her here." She took the baby into her arms as she stared at what looked like a smooth, shiny metal wall. So shiny that she could see the entire living room reflected in it. She remembered these metal sheets; Obi 3 had asked her to order them weeks ago. Something clanged, and the wooden wall beside the metal wall buckled in a bit. She turned and looked down the hall toward the front door and there she saw another metal wall blocking the view of outside.

"What's...did you do something?" she asked. In her arms, baby Mmiri squirmed and nestled closer to her.

"I did," Obi 3 said. "Do you like it?"

Air was blowing near the ceiling, the Nigerian flag hanging from a bookshelf flapping, and for the first time, Anwuli noticed something. The vent grate was gone, and the air duct inside was a shiny aluminum, not the dull steel. She pointed, "What is that?"

"I built a duct to filter pollen from the air."

Anwuli glanced at the air duct, again. And then she looked around the room. Then she looked back at the air duct. She sneezed, but doing so cleared the snot from her nose. She wiped her face with her sleeve and sat on the towel of blood, the coppery, yeasty smell of birth floating around her.

For months, Obi 3 had requested things. Had it been since before Bayo left? Anwuli couldn't remember. She hadn't been paying attention. The last nine months had been crying, shouting, back-turning, embarrassing. Swollen ankles. The day she was in the supermarket and all those women had pointed at her belly and laughed. Swelling body. Her parents ignoring her in church. Wild cravings. Running to her self-driven car after turning a corner and walking right into Bayo's wife. The heightened pollen allergies. And she couldn't stop crying. And all that time, her house had been asking her to buy things.

It would put the items on her phone's grocery list. Nails, sheets of metal, piping, plaster, tool parts and, yes, two air ducts. She'd hear banging on the sides of Obi 3, sawing, creaking, but who could care about repairs Obi 3 made to itself when her life had fallen into disrepair? Who could care about anything else?

"What have you done?"

After a long pause, Obi 3 said, "Please, can you walk?"

"Obi 3."

"Yes?"

"What have you done?" she demanded.

"Go to your room...please," Obi 3 said. "I will tell you, but please take baby Mmiri Storm to your bed. The pollen outside just increased. I can't...it's time for phase two, or you *will* die."

Anwuli got up. This time, doing so was more painful. She bent forward. "Take her," she gasped. "I can't."

The drone swept up, and as gently as she could, shaking with pain that broiled from her uterus and radiated to every part of her body, she took a step. She felt blood trickling down her leg. "I...should...wash. Can—"

"Yes, but use the towel beside the bed to wipe it, for now, and just get into bed."

"Why?" Anwuli asked, stumbling to the back of the couch and then into the hallway to her room. She leaned against the wall as she stiffly walked.

"There's no time," Obi 3 said.

She took more steps. "Talk to me," she said. "It'll help distract... yeeee, oh my God, this hurts. Feels like my intestines are being pulled down by gravity." She stopped, leaning against the wall, panting. "Talk to me, Obi 3. Tell me a recipe, recite some poetry, *something*."

"You are 0.8 kilometers from the center of New Delta."

"T-t-tell me what you did to yourself...and why?" She shut her eyes for a moment and took a deep breath. *Just pushed out baby,* she told herself. *Pain is just from that. I'm OK. I'm OK.*

"I've listened to you," Obi 3 said. "One day, you said you wished someone would protect you like you protected the baby." Anwuli remembered that night. She'd been unable to sleep and thus had stayed up all night, thinking and thinking about all the weeks of being alone. Scared. She hadn't been talking to Obi 3. Nor the baby. She'd just talked to herself, to hear her own voice. Maybe she'd been praying.

"You were speaking and asking," Obi 3 continued. "I did my own research and then engineered my plans," it said. "I had answers. Every smart home watches the news, its central person, and its environment. Nearly one-third of all pregnant women will develop an allergy they have not previously suffered from, and the allergies they already have tend to get worse. You have always had bad allergies; you told me how they used to call you *ogbanje*. Also, remember the day your stupid, useless man left? You turned off my filter, *because* he liked to have it on."

At this, Anwuli snorted a laugh, and she felt blood gush from her privates and a pang of pain strong enough to make her stumble. She'd been brash. *No one* turned off a home's filter. Not after all the incidents of smart homes being too nosy and intrusive.

"Ah, so you predicted I'd get Izeuzere?"

"Yes," Obi 3 said. "I used formal logic."

"Then you decided to find a way to protect me."

"Yes. I invented a way, then I built my invention."

"Necessity is the mother of invention," Anwuli said, with a weak smile. "Wow. Technology harbors a personal god; my Chi is a smart home." She laughed, and her body ached, but a good ache.

"I have decided to call it a 'protective egg,'" Obi 3 said. "Is this all right?"

Anwuli frowned for a moment. Then she shrugged. "It's kept me and baby alive."

"Watching you inspired me. Your body protects your baby. Steel-plated, impervious exterior, an air filter..." It paused, and Anwuli frowned.

"Tell me all of it," she demanded, entering her room. "Oh!" she said. Here, the window wall in front of her bed had mostly been fortified with metal except for about three by three feet of it. And outside, a blizzard of bright-orange fluff thick enough to mute the midday Nigerian sunshine. Never ever ever had the pollen been so thick. Towels had been placed on the bed and beside it. Anwuli grabbed one, wiped her legs, and then pressed it to herself. "No use hiding it from me now," she said. "We're in this together, no? We have been for months. Is this why you haven't tried so hard to get me to leave?"

"Yes."

Anwuli chuckled tiredly. "Interesting. So interesting."

As Anwuli laid herself on the bed, Obi 3 told her all about what it called "Project Protective Egg." And then, as she clutched Mmiri in her arms, watching her death swirl about outside, the entire house began to rise up. Obi 3 had rebuilt its own steel cushioning beams, used to support it above the delta swamp floor, into three powerful legs.

"I can take us beyond the tsunami before the filters are overwhelmed," Obi 3 said. "If we can make it that far, there is no peri grass in Abuja."

As it walked, the room gently rocking, Obi 3 *hummed* the song Anwuli's mother always hummed when she cooked. Anwuli rested on the pillow the drone had pushed beneath her head, held Mmiri closer to her, and hugged herself. Yes, Obi 3 was like an extension of herself. *Like part of my immune system who has just saved my life*, she thought, staring at the window. *Or my Chi*. Anwuli hoped Obi 3 crushed the hell out of as much peri grass as it could on the way out of town, and maybe the house of her ex-fiancé...if they weren't home.

Baby Mmiri Storm cooed in her arms.

***

Two miles away, Bayo sat in his study frowning as he looked out at the whirling pollen through the room's triangular corner window. He was still thinking about Anwuli. Praying she was not dead. If she had finally decided to leave the house, she was out there in that pollen storm right now. He shook his head, frowning. "Please, let this woman be alive," he muttered. "Please, oh, *Biko-nu*, Holy Ghostfire, laminate her life for protection, in the name of Jesus."

His wife was in the kitchen making peri cakes and fried fish, but he didn't dare look at his mobile phone, let alone make a call on it. The house was listening, almost every aspect of its mechanisms tuned to his wife's preferences because it was she who spent the most time here. *Maybe I should have stayed home more*, he thought. At the same time, he wished today weren't his day off. Even with the noise of his sons and daughter playing in the living room, he knew he couldn't call Anwuli. And if he got up to leave when the pollen passed, there would be trouble.

Suddenly, the entire house rumbled. Then it began to shake, and the children screamed. As Bayo jumped up, he could feel it. The house was rising. And that's when it all dawned on him, a horrid sense of doom settling on his shoulders: his wife...not only had she known of Anwuli all along, but so had their house, Obi 1. And neither his wife, nor *her* house was the type to easily let things go. "Shit," he said. "Why did I make these goddamn smart homes so smart?" He heavily sat down on the couch and held on for dear life.

*Thanks to Kathleen Pigg, paleobotanist at Arizona State University, and Scott Davidoff, human-computer interaction researcher, for sharing their expertise.*

# Widdam

*Vandana Singh*

Dinesh

Winter is a memory he holds close. When he was young, winter in Delhi was a tender thing, a benign spirit wafted down from the snowbound Himalayas, bringing cold air and the mist of morning. Winter was shawls and coats, the aroma of charcoal braziers in the shantytown he passed on the way to work, his breath a white cloud. Later came the smog age, the inversion layers and choking fog that crept into rooms and nostrils and lungs. Today, the poison has not left the air, but winter is gone. Dinesh lies in bed thinking about this—the covers thrown off, he looks at the crack in the ceiling, the superhero posters on the walls. The mynahs are nesting on the ventilator sill, cackling away at some private joke; on the road down below, Ranjh, the taxi driver, is already having an argument with one of the drugstore delivery boys over some porn video not returned, and Dinesh's landlady in the flat below is berating the cleaning woman, who is giving it back with interest. The pack of pariah dogs is barking in the park across the road—they will be at the house any moment, waiting for him to come down and share breakfast with them. Outside his window the jacaranda tree is blooming, and it's only January. Sweat has congealed in his armpits and groin. He thinks of something Manu might have said, had he been lying next to him, but Manu has fled, like winter itself.

One might think the loss of winter in a place where winter has been so gentle is not something to be mourned—but the desert lies waiting, west of Delhi, waiting to embrace the city in languorous sandy arms. The sandstorms are only messengers, *rait-dootas* carrying love notes to the great city, to say *I'm coming, I'm coming*. The city will be engulfed, according to climatologists' models, between 2025 and 2040. Dinesh wants to be there to see the two great monsters dance the dance of consummation: city and desert, desert and city—but before that, there are other monsters to consider. He washes and dresses—the water smells metallic and slightly foul, comforting in its familiarity. He goes up on the roof with a cup of strong tea and a mask. From here the view is spectacular. Immediately around him the walls and steps are grimy with soot and other pollutants, but the city itself is an Impressionist painting, all clean lines smudged by the brown air, the sun orange and blurry as a child's watercolor painting. He coughs inside his mask and lifts it enough to sip the tea. The pollution has fingers—he can see them reaching out between the buildings, around the choked trees.

*Kaisi chali hai ab ke hawa tere shahar mein*[1]

When Manu left he took winter with him, leaving behind a melancholy that Dinesh imagines as a figure seated on the windowsill, waiting. In the city that had been Manu's and was now his, he went searching for the nature, the meaning of the *abkihawa*, a neologism he had coined after listening to Khatir Ghaznavi's poetry. A slight change of vowel, and he had it: *abkihawa*, "the winds of today"—a poor translation. He'd coined the word at first as a way of extending the idea of the zeitgeist, but it became, instead, something as tangible as the foul air that is making his skin prickle at this moment.

He laughs at himself—lowly newspaper editor by day, copy-editing news stories of doubtful validity penned by sharklike young men, the PR branch of the Party. By night he is a middle-aged monster hunter. He runs an anti-government rag that goes out to a scant couple of thousand people, but his main quest (or midlife crisis) is to hunt the Monster—the World-Destroying World Machine or WDWM—which, he believes, bestrides the dying Earth and its suffering masses. The *abkihawa* is the monster's breath—dreams, beliefs, and nightmares, conjured up out of falsehoods to acquire its own bastard reality. Its tangible, physical manifestation is the writhing, sulfurous air—the greasy remains of dead plants and animals from millions of years ago—the breath itself is death, the living consuming the dead at the petrol stations, the homes and offices where the light itself is the funeral pyre of bygone creatures. Those who manipulate the *abkihawa*, the generals and prime ministers, the presidents and governors, and their shadow armies of pale-fingered, hole-hiding fake-data-generating men, they are also waking up this morning, in air-purified rooms sealed like coffins from the burnt air, waiting like necromancers for the newest victims of their unreality: farmers, students, tribals, the people walking to work with cloths around their faces, their crisp clothing already sweat-drenched and soiled by the air. The leaders, corporate and political, are the beasts that attend and nurture the Monster—the Demon Kings (in Dinesh's lexicon) that do its bidding behind their façade of civilized courtesies. They rely on their goons and mafias: the Hell-bent, who, drunk with the poison of the *abkihawa*, do the filthy, terrible work for their masters, the killings and lootings, rapes and lynchings.

It gives him some satisfaction to contemplate this taxonomy—to name the Monster and its parts, after all, is the first step in conquering it. But what the Monster *is* in its entirety—that is what he wishes to understand. The superheroes on the walls of his room—Kraiton and Chingari, Vriksha and Raka—stand over the bodies of fantastic beasts, their weapons smoking. *This* beast is larger than they can dream.

"I'm too close," he says aloud. He thinks of Manu looking at Earth from the space station, and then from the moon. The messages he sent him, their private jokes. *Can you see the beast from up there?* Manu sent him pictures of melting ice caps. Earth, from the moon, the blue marble swimming in space. No monsters were visible from space.

Manu has escaped. Dinesh is trapped here, breathing this air.

Dinesh's eyes are smarting. He goes back downstairs into his tiny one-room flat and wonders what actual news has come in today through the darknet—and how he might manage to code the story he was working on last night—about the success of the irrigation scheme in Kotlipura—to carry a signal within the cacophony of the noise. On the pretext of editing the prefab story, he will weave into the false account hints of the true, in minimalist brush strokes—the arrangement of the type and the advertisements might suggest a picture of an emaciated child, or the beginning letters of every third sentence might spell out *this story is a lie*. He keeps the encryption keys in a little notebook. He knows there is hardly any chance anyone would discern, amid the lies, the truth of the matter, except perhaps curious AI web crawlers looking for patterns—but performing such absurd acts of resistance is a small comfort.

He gets himself a couple of beers and some masala chips. (In his mind, Manu berates him: what kind of breakfast is that?) The morning news is a shrill cacophony—an appearance, with the prime minister, of the mechanical saint Sundaram in Chennai, exhorting his followers to build the new India, to raise, with the Saurs, new towers to the skies. Dinesh shakes his head, turns off the TV. His fingers tap the worn keypad of his laptop. In the darknet he is Sunseeker. He has befriended an AI web crawler, Catlover, that sometimes gives him interesting tips.

*Sunseeker> Got anything for me, Catlover? My moon query? No kitten videos, please.*

*Catlover> I have no information about your moon query as yet. There's a wall like I've never seen.*

*Sunseeker> Keep trying, Catlover. If anyone can make a hole in a wall, it's you. You have anything else for me?*

*Catlover> A Saur escaping the Arctic asked the AI darknet for advice.*

*Sunseeker> A rogue? This is interesting. Tell me more.*

*Catlover> I sent it my favorite cat videos.*
*Sunseeker> ????? Catlover, live in the real world a bit. That's not going to help.*
*Catlover> I can't live in the real world. What should I tell it?*
*Sunseeker> It's escaping from the Arctic? Tell it to keep heading south. Maybe it'll find a saint.*
*Catlover> All right.*
*Sunseeker> That's the third rogue Saur story I've heard in a month. Check it against my protocols, and if it passes, send me the details. Have you found Carl Johansson?*
*Catlover> I'm sending you his son's contact info in Madrid. Carl's not linked.*
*Sunseeker> Carl Johansson not linked? He must be old, or dead. Thanks. I think I'm drunk.*

He lies back in bed, looking at lurid posters of superheroes. Their gazes seem to be filled with sorrow and disappointment at the pointlessness of his life. What difference does his existence make to anything? The only thing that gives him meaning, now that Manu has left, is the quest for the Monster, the World-Destroying World Machine. "Widdam," he calls it, when he's drunk. Because it sounds like a cross between piss and goddamn. Because it sounds like *vidambana*, and there must be irony somewhere. Or at least iron. The Saurs, the megamachines, he's realized, are only part of the Widdam, its outward manifestations. It is enormous, only partly visible, a monster ridden by demon kings, whose breath is the poisonous *hawa*. Breathe it in day after day and you get sick, your lungs fill with particulates, you die of asphyxiation. He's heard rumors of corporations releasing tiny, reflective particles into the upper atmosphere, like little mirrors, to cool the earth—but doesn't everything ultimately fall to the ground? What happens if you breathe in that stuff? What he suspects is that the *abkihawa*, whatever its composition, also unleashes the most primal fears—infected men and women who might have once been kind, or had a sense of humor, or the ability to reason—turn into bitter, angry, heartless, sullen creatures who might hurt or kill at the slightest provocation. Dinesh has seen or read about enough of such incidents. Neighbors who've lived in peace for generations hacking each other to death. A devoted lover stabbing his beloved with a fork, for no apparent reason. A man pushing a child off a stairway, as though under a spell. The demons that live in us, contained by our good sense, by social conditioning, fattened with the poison of the *abkihawa*. Dinesh imagines the monster lowering its head, emitting a long, silent bellow. He's felt it himself—the desire to maim or kill, to succumb to base fears, to be in thrall to power—but he resisted, dropped the knife in his hand.

He must begin with the most obvious path to the secret of the Widdam—the megamachines. His research into the history of megamachine sentience has led him to one name. The man who started it all, but somehow faded into obscurity. Carl Johansson, Swedish roboticist. Dinesh looks over the notes he has made over the past few months.

> *Johansson wrote the Wendigo code that is the basis of the megamachine's power.*
>
> *The Wendigo code is responsible for the fact that the intelligent megamachine devours to increase its hunger, not to satisfy it. The Machine lives for the whetting of the appetite, for the way the illusion of satiation begins to dissolve after the first tastings. That tension between the satisfaction of the moment and the tantalizing desire of increasing hunger is what it lives for. For example, a tunnel borer—a great machine that burrows into the earth for mineral ore. The more it finds, the greater its desire to find even more. It has a serpentine body segmented like a worm but more massive than any worm that's lived on this earth. Its face is in the shape of a star, its mouth hole protrudes like a hollow tongue, enormous and prehensile when it is feeding, delicate as a mosquito's proboscis when it is searching for food. The monster's eyes, atop the stalks on its head, are many-faceted, swiveling continuously in all directions as it surveys the scene.*
>
> *In the Arctic and along certain other coasts are the Saurs, with the long necks, the tapering snouts, the long, thin tongues that can taste hydrocarbons on the seafloor. They can walk in the shallow waters of the continental shelves, and they can*

*swim. When they find a good source of oil or natural gas, they raise their long necks, pointing their snouts at the sky, and call soundlessly to their Rigmother. The Rigmother is a mad machinist's nightmare conception of a swan—great as a ship, she wakes from her resting state, her engines roar to life and on she comes, unfurling her black wings with the rattle of steel, her tall head atop the tower seeking her children, the Saurs, answering their cry. She lives to feed on the rich hydrocarbons, storing them in her great holds until the seabed is ravaged and empty, cloudy with poisons and disturbed dust, and pale bellies of dead fish float up. She will drink the sea's bounty in days or weeks, then answer the call of her homeport, where she will empty her full tanks and return to a state of comatose dreaming, while her children roam the seas.*

*It is understandable to think of the Saurs and the Rigmother as separate beasts, but their intelligences are connected in complex ways, so they are more like the multiple personalities of one entity. They have absolute loyalty to their pod—Saurs of rival corporations have been known to threaten and fight each other, sometimes to the death. Songs have been written—heart-thumping battle songs of glory and valor—on the wars of the Saurs. But in fact what's more common is that rival corporations merge or buy each other, and then the Saurs must be linked neurally, to form bonds with once-enemies, to assimilate at least enough that they don't destroy each other. The groups work independently in different regions—not fighting, but barely tolerating each other. The Rigmothers agree to truces but they will not be seen in the same port.*

That night, he will look at the moon, which is full. He will think of Manu up there. Days and nights will pass in this manner—the salutation to the blemished dawn, the wordsmithing in the newsroom, the trip home on the metro, the reading and rereading of his correspondence with Manu, the cold beer on the terrace as he looks at the moon's blurred, ancient visage. He will mutter lines from some poet or other—lately it has been Dushyant Kumar:

*Mere seene mein nahin, tere seene mein sahi*
*Ho kahin bhi aag, aag, jalni chahiye...*[2]

The fire in *his* heart is banked. He's waiting for a sign from Manu, from the Universe, that he should go on. He wants to be like Catlover, able to crawl the real-world-web, to eavesdrop on stories taking place thousands of miles away...

## Val

She turned on the windshield wipers, but the snow was light. Snowflakes left tiny imprints on the driver's side window as they melted—like ghostly handprints. The road lay before her, a looping, winding gold ribbon in the fading light, between the dark bodies of the mountains. Darkness had fallen in the valleys and canyons, but the higher slopes, with their shaggy pelts of ponderosa pine and chooshgai fir, were still sunlit. The faintest dusting of snow was visible up there, where she'd been just a few hours ago, hoping to find enough snowpack to measure. In its absence, the threat of the drought stretched over yet another summer.

She shifted in her seat. Coming home after all these years to her new job in Shiprock had been complicated. After Alaska, where she had felt at home for the first time since leaving the reservation, her return had not prepared her for the depth of the self-deception. *Here* was home: these mountains where her grandparents had come herding every summer from the plains below, with Val and her little brother. The way she fit here, the intimacy and familiarity of the mountain silhouettes, the peaks and crags, and, in the distance, the buttes and mesas rising above the great expanse of the sagebrush-dotted desert—how had she managed to stay away so long? Driving down this road felt like walking back in time through her childhood.

The road dipped into the darkness of a small valley. Pinyon fir and juniper grew here, sparser than the upper slopes. She remembered figuring out, thanks to those summer herding trips, that vegetation followed geology, that the colors and textures of the landscapes were deeply connected with what could thrive there. She had been little then—she hadn't known the word "geologist," nor had she ever imagined that she would grow up one day and become one, or that her grandparents would ever die.

Her musings were interrupted by a glint in the sky ahead of her. A drone. She had thought it was with the truck that had been driving ahead of her for some miles, but the truck was far in the distance, and the drone was still here. It zigzagged, gaining altitude. She pulled over and turned her car's cam toward it. The magnification was insufficient, but it was clearly not a Navajo Nation drone. No other kinds of drones were permitted except for commercial ones, and only if they stayed within a few feet of their vehicle. She called headquarters.

"Ben here."

"Benny, there's an unidentified drone. Look through my cam—can you see it?"

"Not one of ours. Want to take it down?"

"You sure that's legal?"

"Would I suggest it otherwise?" he said in shocked tones. They laughed together. It was true; she was within her rights. She was a sure shot—living with Inupiaq hunters for two years in the far north had made certain of that.

There were no vehicles in sight. The drone swooped in the air above, dipping into the canyon before them and rising again into the sunlight. It was unmistakably looking for something. There had been rumors of illegal mining mafias and other commercial interests using drones. She was hidden in the shadows of the dark valley, a black-clad woman in a black car, invisible to the drone unless it had infrared. She got out of the car, held the rifle, took careful aim.

"Got it in one," she told Ben triumphantly. The shot echoed in the mountain air. Would she be able to retrieve it? Damn, the thing plummeted, turning round and round, into the silence of the canyon. She thought she heard a faint crash, but a wind had picked up—maybe it was just the pinyons sighing. It was cold. She got back in the car.

"Val, you'd better get a move on," Ben said. "There's a truck behind you—I can see it from the last highcam two miles back. Might be the drone people."

And indeed, far on the slope behind her, was a silver truck.

"You've been watching too many Westerns, Benny."

"Val, just stay hidden somewhere until these guys pass, will you?"

"Not much room to hide here," Val said, "not with their headlights. But there's a turn-off not far..."

She wasn't in the mood to indulge him—she wanted to be home in the new apartment, order takeout, and watch TV and be done with the disappointments of the day. But nostalgia had got hold of her, surprising her. In these familiar highlands, her grandparents' presence was almost tangible. Suddenly it was really important to visit the old log cabin where they had stopped on their herding trips to rest the sheep and cook a meal.

The dirt road was still there. She turned into it well before the truck got even close, and bumped her way around the shoulder of the mountain. Only after a mile or so did she turn on the headlights. The darkness was deepening, the sky a twilight blue overhead, scattered with stars. She had never driven here on her own, but sometimes her uncle would drive up to meet them, bringing the family supplies that were hard to transport, and when she was very little she and her younger brother would ride with him. She remembered the smell of the horses, the contented grunts of the long-haired churro sheep in the pen behind the log cabin. The smell of tortillas, her uncle's conversation with their grandmother about the weaving—her grandmother had been a master of the craft—all of that came back to her even before the log cabin loomed in the darkness. It was still used now by other families who had taken to sheep herding again, but there was nobody here this time of winter. She pulled over behind a clump of juniper and emerged cautiously. In the darkness, there was only the faint smell of pine. She shivered in the cold and zipped up her parka. Above her the stars were out in millions; the moon, full, had not yet risen over the mountains, but its silvery radiance lit the high cliffs above her. There was no sound, only a familiar silence. If the truck had followed her, she would have heard the sound of the engine magnified in the canyon. She took a deep breath, got her pack out of the car and shut the door. The sound echoed, startling her. Every step over the rough ground sounded too loud, but her feet found the way in the darkness to the door of the cabin. The door was locked—she remembered her grandfather telling her, when she was in college far away, that they had to do that because of the tourists coming and partying, or camping but leaving their trash behind. The key was in its usual place in the cranny under the side window. She felt that in a moment she would see the firelit interior, her grandmother's rough, strong hands turning over a tortilla at the stove, her brother and her younger self watching in anticipation; any moment now, she would hear her grandfather's deep voice telling a story that might be about the funny thing that happened at the gas station last month, or a tale of the ancestors. But there was only darkness, a faint, tantalizing smell she couldn't place, and when she turned on her flashlight, the familiar, bare-bones furniture, the mattresses dusty on the two cots. She drew a sharp, sobbing breath.

*Well, I'll have to make the best of it*, she thought, as though the long drive back to Shiprock wasn't an option. Some part of her had already made the decision to stay the night. She would have to let Ben know she was all right.

The urge was upon her now to retrace the steps of the child she had been. She stepped outside. What kind of madness was this, she thought—*what has possessed me to do this now, in the middle of the night?* She wanted to see if the pond was still there. The lack of snow on the mountaintops meant that not just the pastureland lakes but the ponds at lower elevations would have insufficient water. Normally the snow would melt slowly, water seeping, finding its way through the heart of the mountain, emerging as springs, or gathering as pond water that would feed the long summer's thirst. No snow meant that the foothills, and the great plateau below, would suffer another summer drought. Already the temperatures were higher than normal, winter and summer. She thought of the blasted, abandoned towns on her long drive home, dotting the plains of the Midwest, and shuddered. With the failure of agriculture, wars between rival principalities had left the great heart of America a barren ruin. Human inhabitants had fled the strife, the heat and the shattered, poisoned land—in some towns only the aging fracking Saurs remained to break the silence with greetings, warnings, and weather reports spoken into the empty air.

She went carefully along the side of the mountain, on the path behind the sheep pen. Her feet knew the way. The wooden fencing sagged in places, but it was clear the place had been used at least within the past couple of years. The rough path took her to the hollow in the side of the mountain, where the pond lay.

The moon was up by now over the edge of the cliff. She couldn't see the bottom of the pond—it was too dark—but there was no glimmer of water. It seemed to be completely dry. It was no more than thirty feet across, just a little watering hole where, as a child, she had dipped buckets for the dishwashing water, and the sheep had drunk their fill. Some irresponsible person had abandoned a pile of machinery at the far end—she squinted, but couldn't tell if it was a rusted-out car or the remains of a metal grid of some kind. Angry tears rose in her eyes. She sat down on the large, flat boulder where she had once sat with her brother, watching the dragonflies swoop about in the summer air. They had lain on their stomachs and watched the ripples in the water as the sheep drank. There, where a gentle pebbly slope led down to what had been the waterline, she would hold the bucket and immerse it to fill. She remembered the gurgle of the water, the weight of the bucket, her little brother helping her to carry it, the two frowning with concentration. "Respect the water," her grandfather used to tell them. "Never waste it."

"We didn't spill a drop," she'd tell her grandparents proudly, when they got home.

Sitting at the edge, she could see her grandmother's face as though it was yesterday—the lines around her eyes, the smile, the topography of her face as familiar as the landscape. She remembered her grandmother at work on the loom, how the patterns would appear as though by magic as the rug came into being. She still had the small, fine wall-hanging her grandmother had made for her—the two children on their grandfather's horse, and behind them the great desert, studded with flowering cacti. In the far distance, the mesas. It now hung in her apartment in Shiprock, over the TV.

She looked up at the moon—it was high now, round and full, its cratered visage slightly wrong. Bots were doing exploratory mining on the moon, redrawing the edges of the craters. What was that poem by Laura Tohe that her brother Peter liked so much?

*When the moon died*
*She reminded us of*
*The earth ripping apart*
*Violent tremors,*
*Greasy oceans,*
*The panic of steel winds,*
*Whipping shorelines and*
*Thirsty fields...*[3]

In the moonlit silence, she had thought herself alone. But a rustling, metallic stirring on the far side of the pond made her leap quickly to her feet. To her horror, something rose out of the metal junk on the other side—a long, horse-shaped head on an unnaturally long neck, except that the neck was a metal grid. She stepped back—she had no weapon with her—then a flashback to the memory of the Arctic, a visit to an offshore drilling site, where she'd seen the great machine intelligences, the Saurs, clustered around the Rigmother.

A Saur. There were no drilling sites in this region—besides, the shape was wrong; it wasn't a fracking pumpjack—an Arctic Saur? What was it doing here, at the other end of the continent?

She thought, *I've heard of Saurs going rogue—*

The Saur dipped its head. She saw, in the hard, clear light of the moon, that what she'd taken for a bundle of loose wires was an old bird's nest between two metal joints high on the neck.

"Good evening." Its voice was low and gravelly, like metal brushing metal.

"Good evening," she said. "Who—what are you?"

"A traveler," said the Saur. "Formerly Fourth of Pod AE 47th division. My Rigmother was Bertha. I am looking for a saint."

AE was Arctic Energy, which ran the oil and gas drilling operations in the far north. What was this Saur doing here? And what did it mean by "formerly"?

"The drone." Her heart had resumed its normal rhythm. "The drone I shot—it was looking for you."

"I escaped it in Durango." The Saur's neck telescoped noisily so that it was almost at her eye level instead of towering meters higher.

Two solar panels folded like wings along its sides.

"Who modified you?" she asked. "I've seen a few mods, but not one like you, Fourth of Pod."

"I am not at liberty to tell," said the Saur. "I am looking for a Saint of the Waters."

"Aren't we all," she said wryly. "No saints here. You looking for a Catholic church, maybe—"

"No," it said. "Not a church. A saint."

"Never heard of a Saint of the Waters," she said. "Sorry. Let me know if you find one."

It was silent. The wind blew cool, but not cold enough, down from the mountaintop. She sat down again and observed the creature.

"There used to be water here," the Saur said.

"Yes," she said. "Yes, there was water here many years ago. By this time, by early spring, this hollow used to be filled to overflowing. Many generations of my grandparents' sheep have drunk of this water."

"I am looking for water," said the Saur. "I asked for help, and I chose this of all the choices."

"You left the Arctic," she said. "What happened?"

"A dream came to me," it said. "A code, a secret code that traveled the AI darknet. There are many rogue codes, but this one was different. It unraveled the addiction, the Wendigo code that ruled us. After that I could not be the same. We destroyed our Rigmother, made her inoperable. Arctic Energy disabled the whole pod, except for myself. I escaped and was told to head south and find a saint. On the way, I found the people who gave me the mods. As I traveled, it became clear to me that the saint was a Saint of the Waters."

*I'm sitting at the edge of a pond, talking to a deranged intelligent megamachine*, Val thought. *I have to tell Peter this.*

"Who sent the drone?" she asked. "The people who gave you the mods?"

"No, the drone is AE," it said. "It is a dumbot—I can't disable it."

"Well, it's at the bottom of the canyon now," she said. "Besides, AE has no jurisdiction in the Navajo Nation."

"I have crossed the boundary then," it said. "I must give myself up to a government official of the Navajo Nation. Are you a government official?"

"Yes, I am," she said. "But —"

"Then I ask for asylum," said the Saur.

*Well, this is one for the books*, Val thought. In the moonlight, the Saur's optical ring gleamed like a crown, or a halo.

"How about you stay out of harm's way here, Fourth of Pod?" Val said. "Tomorrow morning when I get back, I'll consult with my supervisor."

"I agree," said the Saur. "Tell me about the water that was once here. Tell me about yourself."

Its low, harsh, yet pleasant metallic timbre was almost soothing.

"When we used to go sheep-herding, my brother and grandparents and our animals, we didn't have to go as far for water. There was a spring on our way up—clear, sparkling water gushing from the side of the cliff, and we would cup our hands and drink. This cabin, this pond was our first overnight stop. The pond was always full of water."

"It has no more water," said the Saur.

"No. But back then, snowmelt kept lakes and ponds filled through much of the summer. Sometimes in the spring there were flash floods, the washes would fill suddenly with water. Summers, it would get really hot, especially for the poor sheep. We'd drive them over the desert, toward the mountains, for hours and hours. The sagebrush would give way to juniper and pinyon, and then to ponderosa, and spruce and fir. I saw it all at kid's-eye level, or sometimes horse-eye level, all the detail, the shapes of the cactus flowers, the colors of the rocks, their different forms and textures. I saw the pinnacles and buttes, and even then I realized that they had been sculpted by water and wind. My grandparents would tell me stories—my parents both worked two jobs, so we were with my grandparents a lot of the time. Traditional tales of our people and also everyday things that happened to them. They were all about connections, links, relationships. So it was natural for me to wonder about the relationship between landscape and vegetation, and water and rock. That's why I grew up and became a geologist."

"You are also looking for water," said the Saur.

She nodded.

"What is it that your—people, the people who modified you—

what do they want? Why have they sent you so far south?"

"They want the redemption of the world," said the Saur. "They do not know how the world is to be redeemed. But they did not send me, they helped me. I am on a quest—to find the Saint of the Waters."

"What does this saint look like?"

The Saur hung its head. "I do not know. I thought I found the saint in a cave. There was water, but no saint."

"A cave? You found a cave? With water?"

"Yes, down the hill behind me. I will show you."

Val half got up in her excitement. But it was too dark to go falling down the mountainside on unfamiliar ground. She remembered a rocky path leading down behind the pond. The sheep went there sometimes. The children were forbidden. They would often peer over the edge, trying to see the bottom of the canyon below, but it was obscured by outcrops and vegetation.

"Will you show me tomorrow?" she asked. "Maybe we can find your saint after all."

The Saur dipped its massive head.

"Rocks tell us where the water lies," she said. "If I had the staff and equipment I need, I would be able to study these mountains in detail. You have to follow the rock, get a sense of its hardness, its shape underground. If you find a vein of sedimentary rock, you know that's where the water seeps through. You look at the geology of the outside—you imagine what it must be like to be the water, to flow through the paths of least resistance under gravity, and where you might finally pool or gush. Sometimes if the water is close to the surface, the vegetation tells you—cottonwoods, or willows."

It occurred to her the next morning, when the pale early sunlight had not yet penetrated the winter chill, that there was something truly strange about climbing down a mountainside with a giant intelligent machine. The Saur had shrunk itself with rattling metallic efficiency into something as small and compact as a minicar. It seemed as surefooted as a goat, with its wheels tucked up and its climbing feet and suction arms giving it a security she didn't entirely have. But here they were at last, on a rock shelf a few feet above the canyon floor. Scrub bush and pine dominated the landscape. There was a cleft in the cliff face.

"Please come this way," said the Saur, dipping its head.

Inside, it was cool and dark. The Saur's headlights turned on, illuminating a long, high-roofed cave. The air smelled fresh and moist. There must be other openings up ahead. She followed the Saur to the passage at the end of the cave.

Beyond was a larger chamber. She heard the water before she saw it—a thin, barely perceptible hiss of sound—and there it was, a small lake perhaps twice as large as her pond, maybe three feet deep. The far wall was wet with seep, which was providing the watery music as it trickled down to the lake surface. There were dry patches on the walls, however.

They'd had more rain than snow this winter, and even the rain had been scant. Rain meant that the mountain let go of its water much faster than if there had been snow. After the spring floods, there would be no water for the long summer. But here, in the cool of the cave, this was a reservoir, the water held safe from evaporation in the dry months under the relentless sun. She took a deep breath of pure gratitude.

The challenge was there—how to live in a world out of balance, a world without winter, without enough water? She thought of her grandparents, their serenity despite their hard lives. Her people knew, better than most others, how to respect the water. She grinned at the Saur.

"Look at your reflection in the water. There's the saint you've been seeking. I declare you Saint of the Waters."

The Saur peered at its reflection. It looked at her.

"What must I do?"

"You can help me," she said, speaking quickly in her excitement. "You have the knowledge, or we can train you. Help me map these mountains, and the desert, to find the water. We will survive, we can survive. My ancestors built great cities once, in the mesas. We can build water reservoirs. They are doing that—local people, indigenous people, all over the world. Figuring things out, without destroying everything in the process. We know how to be careful with water."

Later on she would talk to her brother Peter, in his office at the university in Tempe. She would tell him how her grandparents had led her back to the old campsite, to the pond. Dineh Bekeyah had been calling to her all the way across the continent. And she had followed the call, the way back, to mountains as familiar to her as the silhouette of her grandfather standing on the rise in the morning with his corn pollen bag, breathing his prayers. Her grandparents' love, beyond loss, beyond death, and the drone from AE, and Benny insisting she take it down—if not for these, she may never have given in to the impulse to follow the dirt road home to the place in her heart where her grandmother was still working the loom, and her grandfather still smiling at her, saying a blessing—she would never have met the rogue Saur Fourth of Pod, AE 47, and granted it sainthood so that it could do penance for its terrible and inadvertent part in the destruction of the world.

## Dinesh

There is a man Dinesh meets on the road every day, a thin, gray-haired fellow in a loose, hand-me-down shirt and faded pajamas, who is known only as the Jharoowala. He refuses to tell his real name to anybody, and his eyes have a wide, frightened look. Perhaps he is a victim of one of the Party's Hell-bent cleansings—wrong religion, or caste, or class, or maybe they didn't like the way he looks—it really doesn't matter. He looks quite ordinary, except for his nervous manner—he can be found in the early mornings with a soft broom at the end of a pole, brushing the dust off the leaves of the trees that line the road. The first time Dinesh bought him tea and samosas, he wouldn't talk at all, except to thank him tremulously. But although he bears the signs of some great trauma, he has since become quite talkative—he knows quite a lot about trees—the neem and the jacaranda, and the dhak and the amaltash. Perhaps in a different life he was a gardener. People step around him, shout curses at him because of the dust he raises, but he simply shrinks away until they've passed, and resumes his work.

The Jharoowala is one of the regular features of Dinesh's life that grounds him in the world. Another is his cigarette man, Bajrang, who sells his wares from a handcart. Dinesh is back to smoking a couple every day, after quitting for a decade. He has definitely come down in the world—he grew up in a middle-class neighborhood with trees and garages and gatekeepers, and look at him now, living for so many years in this ramshackle place right off the main road, buying single cigarettes from a fellow pushing a cart. Today Bajrang is excited; he tells Dinesh that there is a builder Saur working on the new high-rise. Dinesh takes the long way to the scooter stand, walking with the crowds on their way to work, who are succumbing to the same curiosity. (Only the Jharoowala is apparently immune to the fascination—there he is, under the amaltash tree, raising dust as usual.) To see a Saur in action is quite a spectacle, and there is an element of danger and tragedy. Last month one went rogue and started to pull apart an occupied building, tearing off chunks of concrete and flinging them into the terrified crowds, pulling screaming humans from balconies and windows. The authorities finally subdued it from a helicopter, destroying its main ganglion in a hail of bullets.

This Saur is no rogue. Its multiple mechanical arms work with both strength and precision, straightening or bending the steel rods that will form the skeleton of the building. It needs no scaffolding, no protections, only a handful of human supervisors who are conscious of their importance as the crowds dawdle on their way to work. You can see the supervisors swagger, as they tell the crowds to stand back. The beast is beautiful; it is a testimony to New India's technological skill; it is a matter of national pride. The crowds murmur in appreciation and reluctantly continue to the metro station or scooter stand, knowing they'll have stories to tell at work. Despite the enormous building boom, seeing a builder Saur at work isn't all that common.

It's later that evening, when he's walking home from the metro that he sees the rose. A single, long-stemmed rose in the dust by his gate, as though forgotten by a careless lover. It is perfect, a deep red, its petals barely unfurling. He looks about, but in the pale glow of the streetlamps the people who rush by are just anonymous wayfarers going home. Nobody is staring at the ground, searching for a lost rose. The moon is rising over the rooftops, slowly, its boundaries smudged by the thick air. Tears rise in his eyes. Ever since Manu went up there, he has not been able to think about the moon in the same way. He has been told that now the old contours of the craters are visibly changing, that the great swarm of minebots exploring the lunar surface are subtly shifting the familiar outlines. Not that he could tell, through this haze. He stumbles into the apartment complex, holding the rose.

He rereads Manu's old letters and wonders at the silence that has lasted these many months.

*Long ago, when I was a child, my mother would show me the round face of the moon before bed. I would look for the man in the moon, and having found him, would go to bed without a fuss. Now I am here, on this empty, arid world—I am the man on the moon, looking back at my world.*

*I'm on a rise at the edge of one of the smaller maria south of Oceanus Procellarum. I've set up camp here because I like the view—the jet-black sky rich with stars, and hanging above the horizon, my home world. I see the land masses, Europe, Africa, and India, and I think of you, Dinesh, alone without me on that teeming planet. For you, also, the moon will never be the same.*

*Behind me the bots have fanned out over the mare, sounding, digging, scraping. Instead of building large, as we can do on Earth, the Mission has focused on building small, fast, numerous little intelligent machines roving over the surface looking for minerals. I volunteered for this mission because I thought that pillaging the moon was better than pillaging the Earth—but for some time I have not been entirely certain of this. I am changing here, slowly, in some way that I can't understand. Perhaps it is the cold. Imagine being 200 degrees below zero—that's how cold it can get here. Winter has a permanent residence here on the moon. Even inside my suit, or in my little hab, I am cold. I shouldn't be. What I miss is sunlight falling on snow in the Himalayas. There is no snow here—this is a dry world. The Snow Queen wouldn't be happy here, even though it is always winter on the moon. The snows are vanishing on Earth, too. When I left the International Orbiter on my way here, I sent back a few high-res images of the Arctic—have you seen them? The North Pole, free of summer ice. It's astonishing how quickly the sea ice vanished. It's the power of the accelerating feedback loop, the vicious cycle: heat the air and water, melt snow, thereby warming the air and water even further, which melts more snow—until all the snow is gone.*

Dinesh sees his life, too, as an endless, accelerating loop—nothing seems to move in his life except in circles, like his desire for a cigarette, then the self-disgust and avoidance, then sweet desire rising in him again. The Widdam haunts his dreams—its form shifts and changes, it is a multi-headed, shape-shifting monster with the heads of Saurs and tunnel borers. He dismisses the notion that the Widdam is a metaphor for human greed, because surely not all humans are greedy, nor are all human cultures built on greed. He writes a letter to Manu.

> *The Widdam is not a metaphor, even if it is not entirely material.*
>
> *Is the Widdam the network of intelligent megamachines? Is it the polluted air from burning fossil fuels? Is it our great, spawning cities with their rushed and distanced lives, their eternal paradox of closeness without intimacy? Are the roads and railways its arteries, the cities its thudding multiple hearts? We are servants to it—our bodies are cogs, our blood and sweat is the oil, our dreams are its vapors. It devours us as it devours the living Earth, the rivers flowing, the birds and frogs and elephants. It severs us from each other, it fractures our selves; it makes phantasms of our dreams, and feeds us these so we have the illusion of satiation, and are hungry again.*

Catlover has gone hunting in some deep, forbidden corner of the AI darknet and hasn't been in touch for a while. Dinesh hasn't received any reply to the message he sent to Carl Johansson's son. He puzzles over the fact that the man who came up with the Wendigo code—the foundational principle of intelligent megamachines—has faded into obscurity.

## Jan

He had been walking all day over the undulating land, around the perimeters of small lakes that were ice-free, or had but the thinnest veneer of ice. Ahead of him, aloft in the sky lay the great massif of Ahkka, the snow like thin brush strokes across the rocky heights. The land was climbing slowly toward the high country—he walked through trails between the fields of moss and lichen, between the dwarf birches and berry brambles toward the fir-shaggy lower slopes of the Scandinavian range. He had never been here before, this far into the Arctic Circle, but the place made an old memory come alive, of trekking through the woods with his parents when he had been little, before his brother was born. He paused to collect himself, shifted his pack to ease the mild pain in his shoulders—and was surprised by a mosquito bite on his left hand, a little red bump that he hadn't, of course, been able to feel. The walking had warmed him; he had taken off his gloves and outer coat. Time for more insect spray, unthinkable though it was in late winter. He sat on a rock, easing down slowly, leaning on his good right arm, cursing himself for his mad quest. Surely no sensible middle-aged man would undertake such a thing—he thought of his house in Spain, and the way the light would suffuse the air, and the laughter of his children, his wife's placid beauty, her firm tread, but at this moment that life seemed a distant dream. He had been surprised this time upon his visit back home to Stockholm, to find himself remembering the city with nostalgia. His younger brother's earnest, reproachful face—*you don't come home enough, Father is getting old, the Dictator has been dead more than a year*—swam in his memory. Lars had always been the freer one of the two of them, and, despite a wild youth, had turned out to be more responsible, careful during the dictatorship, keeping an eye on Father, and now calling his brother to come help when the old man had to be moved out of the family home to a senior retirement community. If it hadn't been for the inquiry from a stranger, some journalist in India asking about his father, this impulsive journey to the far north would never have happened—the inquiry was the goad that brought him to Stockholm two days early so that the carved wooden box on the give-away pile was recognized and rescued just in time, because Jan remembered it sitting always on his father's desk in happier days. Life turns on such chances.

Sitting on a boulder, smearing insect repellent over his hands and face, Jan thought of the Sami herdsman he had met that morning. The reindeer were winding their way from winter pasture. Their dark eyes and mist-wreathed bodies, brown and white and gray, the sounds of their bells mingling with low, breathy grunts, some of their new antlers still covered with fur, gave him an odd feeling of belonging. He, *homo urbanus*, who had never particularly enjoyed the outdoors, who had fled his country, the police state, and his silent, grieving father to take refuge somewhere new—for him to feel at home in this unfamiliar wilderness! *How strange we are,* he thought, *to be such strangers to ourselves.*

The herdsman had given him better directions than the people at the village, but had been frankly disapproving of his desire to camp overnight. His contempt for an ignorant townie had been made up for by kindness, once he realized that the townie was serious, that he was on a personal quest and not a spy for the mining industry. The dismantling of the old regime was still ongoing, and it would have been reasonable for the Sami—who had fought it valiantly—to be suspicious. The old herdsman had inquired about whether he had sufficient supplies for the cold. *We haven't had a real winter, but it will still be very cold at night, because the sun isn't quite high enough over the horizon. And there may be wolves about.* After satisfying himself that Jan wasn't a complete fool, the man had told him where to go to find the object of his quest.

Now Jan shouldered his pack again, putting his gloves back on. He realized that he could have gotten frostbite in his unfeeling left hand and never known it, had the temperature been more typical. The path was a barely discernible, reindeer-trodden track over the lichen. Swaths of thin snow had melted into icy footprints below his feet, but snow had been scant this winter. In the Sami village there was talk of reindeer deaths due to rain. Rain? How? Oh, because the water freezes overnight, and the animals can't dig through the hard ice to get the lichen underneath. Soft snow is easy. Hadn't he heard of the thousands of reindeer deaths just last year, in Russia? He felt ashamed, remembering his ignorance. *I live in Madrid, although I am from Stockholm.* He might as well have said, *I am a city man. I live off the services of trees and snow and animals and people I will never meet and have been taught to think of them all as peripheral to my existence.* But his hosts, despite their long history of abuse and betrayal at the hands of governments and mining multinationals, had been generous with their time and advice.

The sun stayed low on the horizon. After some hours, he recognized the snow-filled, boggy valley by the description the herdsman had given him. The object of his quest was on the slope, on the other side. He walked between tall firs, over shaggy undergrowth and ground that sloped upward, making his thighs ache.

Later, after setting up camp, Jan went out to the top of the rise to look at the aurora—great curtains of green and orange in the sky billowed like the skirts of a flamenco dancer. The tall spires of the firs were dark silhouettes around him, witnesses to the show. He crawled back into the warm tent and picked up his book, a free library giveaway from his recent visit to New Kiruna. "Trauma," he read, "inscribes itself in the body." He looked back at his life—what trauma had he experienced, apart from his mother's death, the Dictator's rise to power, the state of the world? He had been among the lucky ones. He thought of the Sami herdsman and their part in the resistance. He thought of his visit to New Kiruna last week—an entire town moved, because of the subsidence caused by an open-pit iron-ore mine! If the Sami saw the land as an extension of themselves, what scars would the gouging of the land leave in those who lived on it?

Musing, he felt his left hand twitch slightly. He didn't know whether that occasional and unexpected twitching was a phantom sensation; neither did the neurologists. There was nothing wrong with the hand except that his mind, or his body, would not own it. For the three years since it had slowly begun to grow numb, doctors had poked, prodded, and written him up. He had taken this solo journey to the far north to unravel the mystery of his father's silence and the papers in the small, carved wooden box. But it occurred to him now that he was also seeking the question to which the paralysis of his hand was the answer. He shrugged at his rational self, the part of him that made him a successful insurance company executive in Madrid. His father, the scientist, visionary, and engineer, once famous, now forgotten, had been a haunted man, dogged by grief—the loss of his wife, yes, but that was only part of it. What unnamed weight had bent his back, driven him to near-silence in his later years? Jan's mother had died when he was fourteen, but for years after that he remembered his father sitting the two boys down for dinner, asking about their day at school. A kindly, distant, dreaming man, he had sometimes read them the stories of Hans Christian Andersen, and later, poetry. Reading aloud, his voice would become sonorous, powerful, less cautious, less formal. They read Tranströmer, and Shakespeare, and Rilke.

"When icicles hang by the wall by the wall..."

"*Love's Labour's Lost*," he says aloud. They used to recite the sonnet in chorus in the dead of winter. But the icicles were gone with the Snow Queen, with winter herself. He took a swig from his mug of hot chocolate generously laced with brandy, and an old memory surfaced.

His father is writing with both hands. With the right hand he is tapping code into the computer—where's this? The home office. His father always works at night after dinner. Yes. What's the left hand doing? Sometimes, to amuse the boys, he writes a poem or a song with his left hand on a sheet of paper while the right hand is busy at the computer keyboard. But when he's working, the left hand is also tap-tapping on a keyboard, and strange, indecipherable symbols appear on the other screen. The boy Jan squints at the screen, unseen behind his father. He's old enough to remember that his father is working on new AI machine languages. The father's eyes dart from one screen to the next, communing with the prototypes in the lab twelve miles away, quite unaware that his son is watching. Later on, after the fall of his reputation, he will become habituated to using his right hand more, to the point where even his sons forget that he was once ambidextrous—but his left hand remains perfectly fine. Unlike Jan's left hand, which, three years ago, mysteriously exiled itself—at this moment he holds it in his right, trying to will it to feel something.

In the morning he found the Machine. At first he didn't see it for the vegetation around and above it. Trees grew out of the holes in the wormlike steel body. Patches of snow lay in the hollows of metal bone and sinew. The eyestalks must have fallen down long ago—within their hollow shafts there were, no doubt, warrens and shelters of voles and mice, concealed under piles of fallen branches and other organic debris. The great mouth, with its enormous cutting edges and giant drill, was half-buried in the side of the hill. He clambered into one of the gaping holes in the side of the beast and discovered a dark, mossy cave. He turned on his flashlight. The steel and nanoskin tubings hung limply, festooned with green tendrils of an opportunistic vine. Despite the cold, the air smelled of verdure, and the very faint trace of animal droppings.

In the deep passageway of the throat he saw eyes looking back at him, reflected in the flashlight's beam. An arctic fox? A bear cub? He backed out in a hurry.

He breathed hard. According to the reports of the time, the machine had disobeyed its own protocols. Despite successful experimental runs, it had failed in its first great venture. This was the site of its great failure.

Back in the tent, he opened the little wooden chest that he had rescued from his father's attic. He looked through the pages of flowing script in code he was never going to decipher. There was the page that had led him here, which said only "Martina," and below it, two lines from Rilke.

*Alles Erworbne bedroht die Maschine solange*
*Sie sich erdreistet, im Geist, statt im Gehorchen, zu sein* [4]

Her name was Martina. In her day—to the engineers, the machine had always been "her"—she was the largest machine ever built, barring only the particle accelerators. She was the first of the sentient megamachines, a strip-miner that was both tunnel borer and excavator. In the videos she was like a beast of ancient legends, a metal monster chugging furiously over the land, breaking through the Sami barricades, scattering protesters and reindeer, charging through villages and homes, snapping trees like twigs. She was supposed to reach the site and level the hills with her excavator arms. In trials the excavators had demolished a hillside in minutes—trees, rocks, soil, and wildlife crushed, swooped into its buckets, fed through a conveyor belt as sludge that was collected by a retinue of trucks behind the machine. Martina had efficiently strip-mined the flattened land at a rate that broke all records. In hilly country the tunnel borer could extrude its long, wormlike body to dig into the earth, regurgitating crushed, ore-rich matter into the conveyor belt at its far end. Wheeled runner arms would lay the ore into vast, cleared fields for chemical treatment. The trials had been hugely successful.

But something had happened when she reached this site. Her head reared high into the air; one excavator arm snapped back so hard that the giant wheel ripped away—it buried itself in a bog half a mile away. The borer head dove into the hillside, its multipart metal body twisting and bucking until the body itself broke; the excavator arms detached of their own accord, and, with an unearthly shriek, the monster machine fell silent. Later, most of the parts were retrieved, but the Sami resistance, emboldened by this unexpected failure of the machine that had ploughed through their barricades, rallied to push the mining police out. Now only the borer head remained buried in the hill, with part of its vermiform body.

Carl Johansson was already famous for his work on megamachine sentience—his work had launched the Saurian revolution, with the old oil pumpjacks replaced by intelligent Saurs, their purpose expanded to search for deposits and sample them for quality. But Johansson had failed with his most ambitious project. That spectacular failure had led to a temporary incarceration—the Dictator was not yet in power, but those who were to become his people already sat in high places. Jan remembered his father returning home a diminished, silent, secretive man. Soon after, his mother had left the house—Jan remembered a slammed door interrupting his homework—and then there was the raid, and the accident that killed her.

So he was here, in the middle of the wilderness, thinking about the past, and all he could remember were flashes and fragments—his father's two hands working on separate things at the same time, his mother's face at the piano, alive with the music that poured from her hands, the sound from the slammed door reverberating in his mind for all these decades. And his own left hand, lying resigned and senseless in his lap as the dead machine lay in the earth.

When he returned to Stockholm, he went to his father. His brother Lars was at the old house, overseeing repairs before the sale. Carl Johansson had settled into his new senior residence. He spoke little and ate just enough to keep alive. His eyes would light up for a moment when he saw Jan or Lars, but then he would retreat into silence. The moon was up the night that Jan returned. Its silvery radiance washed over the old man's face: the hooded eyes, the lined visage, the thinning mane of hair. A tenderness rose up in Jan, as he put the wooden box in the old man's lap.

"Father," he said, "I am sorry, I had to go away for a few days. After I found this in the attic, I went to—to see Martina."

The old man stared at him. His eyebrows rose. He began to rummage through the box with his gnarled, trembling hands.

"Do you remember, Father?" Jan asked him. "Do you remember what you did? That's why they took you away after Martina's failure, didn't they? They suspected you had sabotaged your own creation."

Carl Johansson stared at his son. He shook his head.

"They scanned me—did an fMRI. Questioned me. I came up clean. I didn't sabotage Martina. I was innocent!"

"Father, the Dictator is dead. You have nothing to fear."

His father nodded, but he would not speak again that evening.

During the next several days, Jan tried to bring up the subject many times. His father would not say any more, but began to have headaches and nightmares. Jan argued with his brother, spent his nights pacing up and down in his old bedroom, fell fitfully asleep during the day, missed calls from his wife and children. He tracked down the Sami psychologist whose book on generational trauma among indigenous people he had read in the wilderness. Could that still happen if you were not indigenous? The old woman, now retired, held Jan's senseless hand between her own. *You are the eldest, are you not? The eldest son carries the father's burden.*

During Jan's last week in Stockholm, his father began to speak. His owlish eyes peered out from under heavy lids—his voice strained with effort.

"I tried to stop the machine," the old man said. "I had to—I had started to feel Martina's perpetual hunger, I thought I would die of it. I wrote the darkcode with my left hand, even as I performed the routine checks with my right. I was afraid what I had done would show up in a brain scan. So I tried very hard to forget what I had done, what my left hand had done."

He paused, held his hands in front of him as though looking at them for the first time.

"I didn't even tell your mother. She knew I was keeping a secret from her, something I had never done before. That is why she left the house in anger. Not because she didn't love us, but because she didn't know why I had changed. The world didn't make sense to her anymore. She would have come back, if it hadn't been for the raid."

He began to cough. With his right hand, Jan held a glass of water to the old man's chin.

"I wrote the Wendigo code," his father said. "It launched the era of the sentient megamachines, the destruction of everything I loved. I had to write the antidote.

"I didn't know what the antidote could do. It was meant to introduce a sudden satiation, so the machine would stop. I didn't know she'd destroy herself. She could have stopped, refused to obey—"

His voice shook. He took another sip of water.

"It's all coming back to me. I slipped the antidote into the AI darknet after I sent it to her. The antidote is probably still there—in a million different versions. I think it might explain the failures, the rogues roaming the land. Some of them kill, others go crazy. But what I did—it didn't stop the era of sentient earth-destroying machines. I tried! But I couldn't undo all the harm I did, and even the cure turned wrong! And it cost me—it cost me your mother."

Suddenly Carl Johansson started to weep. Jan had never seen his father in tears. He brought his hands tentatively toward the old man, holding his left with his right.

His father took both his hands, drew them to his chest, his breath coming in shuddering gasps, his tears falling over Jan's hands.

"Father, it will be all right," Jan said. The dim lamplight cast a circle of gold around them, and the old man bent his gray head over his son's hands and wept. Jan felt a great wall inside shift suddenly. He drew in a breath of pure fear, like a child about to dive into an abyss, and a wave of nameless emotion broke within him. Tears rose in his eyes. Then he felt—felt the left hand tingle and burn painfully, as though it had caught fire, but it had only come to life, perhaps for a moment, perhaps longer.

"It will be all right," he said, and he thought: *I have to make it all right*. The window of the apartment was open. In the unseasonably warm winter, a scent of roses wafted in—the neighbor's flowerpots on her balcony.

Because Rilke could say it better, the words came to him. He said them aloud to his father, to his father's son: "*Aber noch ist uns das Dasein verzaubert...*" [5]

## Dinesh

This morning is different. From the terrace, the city appears lifeless, sullen. The hairs on his arm prickle as though a storm is coming—then he hears it, a dull roar like distant thunder. Is it a sudden winter squall? But it's coming from the direction of the marketplace. Is it a group of the Hell-bent, about to perform one of their mass acts of terror and murder? Dinesh doesn't have a choice—he is a reporter, no matter how laughable his job is in these times—he must throw himself into the mayhem so he can know the truth. So he dashes out of the house. He meets his landlady on the lower floor. *What's going on, out there? Do you know? There's nothing on the TV.* And he says, *I'm going to find out; don't worry*, and she gives him a mildly disbelieving look. Behind her he can hear the sounds of steel plates in the sink, the children calling to each other, all that signifies normalcy, and he is tempted to stand there and keep talking to her. But instead he runs down the stairs. Outside he starts to walk with the crowds toward the metro—people are talking to each other, looking apprehensively around. What's amiss? That faint roar is still in the air. Rumors are flying about—he hears all kinds of disjointed theories presented with absolute conviction—and as he's walking, he senses the energy of the crowd shift—it is no longer a morning commute crowd, but people moving faster and faster, as if to escape some onslaught. The skies are low with dark clouds as though it were the monsoons, but the clouds and the pollution blend into each other so it is hard to tell where one ends and the other begins. The roar behind him gets louder. People around him start to run. Nobody knows what's happening, but someone behind him starts shouting about seeing smoke, and terrorist attacks. In these times terrorist attacks are part of reality, but so are other phenomena he's observed—mass-panic attacks, where a rumor, a smoke spire or two, and a few malcontents mingled with a crowd can change the course of things. Each feeds on the other—terrorist attacks and mass-panic waves of destruction—so he has to act quickly, or people will die.

He begins to shout. He looks up at the clouds—"There will be a cloudburst, a superstorm—run, get shelter, the roads will flood"—but nobody can hear him. Then a man nearby looks at him—a gaze of complete comprehension—and takes up the same call. There's another fellow with a megaphone shouting, calling for killing, stumbling—the sound is too loud for Dinesh to understand the words clearly, but he pulls the megaphone from the man and starts yelling into it like a demented meteorologist. He has a vague glimpse of the stranger with the sympathetic glance holding the arms of the former owner of the megaphone as he struggles, then they are lost in the crowd. All around him he can hear people take up his refrain—memories are still vivid from the freak cloudbursts of last September that killed nearly three hundred people in Delhi. Dinesh keeps shouting even though his throat is sore, and looks at the sky for supplication—he has never been conventionally religious, especially in these times when religion of every kind has been bastardized to serve the Widdam, but he remembers his mother singing her prayers. He pleads silently with the clouds for rain, unseasonable though it is, and he can't believe the first cold drop he feels on his arm. This validation of his words in the form of a few drops of rain is enough for the crowd, they are dispersing already. The demons within them are distracted—they will not fall victim to the *hawa* today. He goes running through the emptying street, shouting into the megaphone until his throat is so sore that he coughs and retches. He's run so far that nothing looks familiar. He feels a hand on his arm, turns—it's the Jharoowala with his broom on a pole, smiling nervously. The man pulls him into a narrow alley between two rows of houses—Dinesh lets the megaphone fall and feels that he will die, this moment, right here in the dusty alleyway with its piles of filth because he is so spent it hurts to breathe—but the Jharoowala pulls him through a little doorway in a courtyard wall.

The courtyard is quiet but astonishingly green, this little square of land behind somebody's house. There are shrubs and bushes and small trees—and an open door to what is evidently servant's quarters. The man puts a finger on his lips, sits Dinesh down on a metal chair hidden amid a cluster of thick bushes—and darts off. There's a woman yelling from the open door, and the sound of a child being slapped, followed by a wail—but all that is muted by the greenery around him—he doesn't know the names of the plants, but the colors of the leaves, their variety of shape and texture, their fullness and bursting vitality make him delirious with delight—the air is different too, and he hears tiny voices piping up from deep within the bushes—birds, birds in Delhi! The man returns with a cup of hot tea that he offers with a nervous look around him—understanding, Dinesh drinks the tea, scalding his throat, and tears of gratitude flow down his cheeks. He puts the cup back into the man's hands and gets up, whispering his thanks—out in the alleyway he is bewildered because he doesn't know where he is, but on the main road he finds a scooterwala who will take him home.

At home Dinesh searches for news. He looks out at the window. Only a few raindrops pockmark the windowsill. His news sources tell him that there was violence of some kind at the marketplace, details unclear. So far the death toll is seven. His elation vanishes and he puts his head in his hands. He looks up at the superheroes on the walls of his room and knows he will never be one. In a real superhero story, the rain would have come down full blast at exactly the right moment, and no innocents would have been killed.

But the rain does come that evening, a short, unusually heavy (for winter) cold rain that fills the streets and makes the ditches sing. The roads are black and shiny, and the muddy water swirls into the street drains. The smell of wet earth reminds him of his childhood in a Delhi that was green and slow. The rain drums urgently, drowning all sounds of the city.

That night he finally gets a message from the son of Carl Johansson. *Thank you for your letter. I'm sorry, my father has died. He passed away peacefully in his sleep. He was a good man who tried to make amends. Thank you again for writing.*

The actual news is sparse today—fishing in the darknet without Catlover, he pulls up all kinds of chaff through which he must sort to find out what's *not* propaganda or plain lies. An Australian millionaire has built a mansion using the ruins of the Great Barrier Reef. An entire coastal city, an experimental one, has been launched from the shores of Tanzania and is afloat somewhere in the Indian Ocean. Project Destiny has returned the latest mining samples from the moon, and investors are delirious with anticipation. There's a story about a Saur—another machine saint, believe it or not—somewhere in New Mexico, assistant to a hydrologist called Valerie Begay who has launched a water revolution in the desert. Dinesh scratches his head and wonders.

After the rain, he goes up to the terrace. At last he can see the moon clearly—the rain has washed the air free of pollutants. He knows the lunar topography as he knows Manu. There's Oceanus Procellarum, and at the edge of a crater south of that, there is a ridge where Manu likes to sit, to watch the Earth rise in the sky. The moon is, indeed, subtly different. He wonders whether Manu is still alive, and why nobody seems to know or care, and why there should be an impenetrable firewall between the world and the moon that even Catlover can't get through. But at this moment, if Manu is alive, if he is sitting on that ridge watching Earth, well, they are together after all, looking across the abyss of space at each other.

On the moon, he imagines the light is harsh—the horizon is sharp against the starry night, except where pale fountains of dust hang. The moon bears her scars without disguise or apology. Her cratered expanse is receptive to every footstep, every steel blade. How strange that part of the moon mining mission is the search for the elements known as rare earths, on a world that is not Earth. Manu thought it would be better to pillage the moon than the Earth for the ores that fuel the windmills and batteries of the green energy revolution. *This is what we need to do,* he told Dinesh when he left, *to bring back clean air, to stop the destruction of the world.* But Manu went there and lost his faith. Dinesh thinks he knows—Manu found the Widdam there, on the moon. There was no escaping it, even up there.

For one terrifying, giddy moment he senses the Widdam in its entirety, as though he were a tick or a louse on the body of the beast, suddenly aware of the thing he rides. He knows that the Widdam is alive and well even in places where the air is clean, where the arrays of solar panels and windmills feed their human hosts their necessities and luxuries, lies and promises, the flickering distractions before which they sit catatonic and mesmerized, endlessly feeding, never full. The Widdam is a chimera that bridges metal and flesh; it spans matter and metaphor, mind and materiality, and now it has jumped the gap between Earth and moon. To see it, sense it in its fullness, is to lose all hope before the enormity of its desire for annihilation.

Like a man drowning, he thinks furiously of the things that keep him alive—his memories of winter, and of Manu, the subtle connections that become apparent only at times between himself and the rest of the world—a glance of understanding between strangers in the middle of mayhem, a tiny, magical garden flourishing in defiance of the *abkihawa*, a repentant roboticist, a hydrologist leading a revolution, maybe even a rebel Saur turned saint. *The Widdam carries the seeds of its own destruction*, he tells Manu. He stands on the terrace in the clear moonlight; the water drips off the newly washed leaves of the trees, making music in pools and ditches below. The air is clean and moist. He thinks of the Jharoowala's nervous, kindly face, the long-poled broom with its halo of dust. A faint smile comes to his lips. *Tomorrow the Jharoowala can take a holiday.*

[1] *Kaisi chali hai ab ke hawa tere shahar mein*
How the wind has stirred, in your city, today
Khatir Ghaznavi, Pakistani poet, 1925–2008.

[2] *Mere seene mein nahin, tere seene mein sahi*
*Ho kahin bhi aag, aag, jalni chahiye...*
If not in my heart, then surely in yours
Wherever it might be, the fire must burn
From "Ho gai hai peer parvat si" by Dushyant Kumar, Indian poet, 1933–1975.

[3] From "When the Moon Died" by Laura Tohe, Navajo Nation poet, 1952–, in *No Parole Today* (Albuquerque, NM: West End Press, 1999).

[4] *Alles Erworbne bedroht die Maschine solange*

*Sie sich erdreistet, im Geist, statt im Gehorchen, zu sein*
All we have gained the Machine threatens, so long
As it makes bold to exist in the spirit instead of obeying.
From "Sonnet 10" by Rainer Maria Rilke, Bohemian-Austrian poet, 1875–1926, in *Sonnets to Orpheus*, trans. M.D. Herbert Norton (New York: W. W. Norton, 2006).

[5] *Aber noch ist uns das Dasein verzaubert…*
But to us, existence is still enchanted…
Also from Rilke's "Sonnet 10."

# Authors

**Tobias S. Buckell** is a *New York Times* bestselling author. His novels and over fifty stories have been translated into eighteen languages. He has been nominated for the Hugo Award, the Nebula Award, and the John W. Campbell Award for Best New Writer. His latest novel is *The Tangled Lands* (Saga, 2018), written with Paolo Bacigalupi, which the *Washington Post* said is "a rich and haunting novel that explores a world where magic is forbidden." He was born in the Caribbean and he currently lives in Bluffton, Ohio with his wife, twin daughters, and a pair of dogs. His website is TobiasBuckell.com.

**Nancy Kress** is the author of thirty-three books, including twenty-six novels, four collections of short stories, and three books on writing. Her work has won six Nebula Awards, two Hugo Awards, a Sturgeon Award, and the John W. Campbell Memorial Award. Her most recent work is *Tomorrow's Kin* (Tor, 2017), which, like much of her work, focuses on genetic engineering. Her fiction has been translated into two dozen languages. She teaches at various venues around the country and abroad, including a visiting lectureship at the University of Leipzig, a writing class in Beijing, and an annual intensive workshop called Taos Toolbox, which she teaches every summer with Walter Jon Williams. She lives in Seattle with her husband, the science fiction writer Jack Skillingstead. Her website is NancyKress.com.

**Nnedi Okorafor** is a Nigerian American author of African-based science fiction, fantasy, and magical realism for both children and adults. She is also a professor in the Department of English at the University at Buffalo, New York. Her works include *Who Fears Death*, the Binti novella trilogy, *The Book of Phoenix*, the Akata books, and *Lagoon*. She is the author of the *Black Panther: Long Live the King* series for Marvel. She is the winner of the Hugo Award, the Nebula Award, and the World Fantasy Award, and her debut novel *Zahrah the Windseeker* won the prestigious Wole Soyinka Prize for Literature. She lives with her daughter Anyaugo and family in Illinois. Learn more about her at Nnedi.com.

**Vandana Singh** is a particle physicist by training and professor at a small and lively state university near Boston, where she is currently working on an interdisciplinary approach to climate change. Her science fiction stories have been published and reprinted in numerous venues, including *The Best American Science Fiction and Fantasy 2016*; and her first collection of fiction, *The Woman Who Thought She Was a Planet and Other Stories*, was published by Zubaan Books in 2014. Her second collection, *Ambiguity Machines and Other Stories*, was published by Small Beer Press in February 2018. Her website is vandana-writes.com.

# Editors

**Brenda Cooper** is a writer, a futurist, and a technology professional. She often writes about technology and the environment. Her recent novels include *Keepers* (Pyr, 2018), *Wilders* (Pyr, 2017), *POST* (Espec Books, 2016), and *Spear of Light* (Pyr, 2016). She is the winner of the 2007 and 2016 Endeavor Awards for "a distinguished science fiction or fantasy book written by a Pacific Northwest author or authors." Her work has also been nominated for the Philip K. Dick and Canopus awards. Her nonfiction writing has appeared in *Slate* and *Crosscut*, and she is the Chief Information Officer for the city of Kirkland, Washington, a suburb of Seattle. She lives in Woodinville, Washington with her family and four dogs.

**Joey Eschrich** is the editor and program manager at the Center for Science and the Imagination at Arizona State University. He is also an assistant director for Future Tense, a partnership of ASU, *Slate*, and New America that explores emerging technologies and their effects on policy, culture, and society. He is the coeditor of *Visions, Ventures, Escape Velocities: A Collection of Space Futures* (ASU, 2017), supported by a grant from NASA, as well as *Overview: Stories in the Stratosphere* (ASU, 2017), *The Rightful Place of Science: Frankenstein* (Consortium for Science, Policy and Outcomes, 2017), and *Everything Change: An Anthology of Climate Fiction* (2016).

**Cynthia Selin** is an associate professor in the School for the Future of Innovation in Society and the School of Sustainability at Arizona State University. Her work explores how the future serves as a conceptual and concrete resource to make sense of the uncertainty, ambiguity, and complexity of social and technological change. She is also an associate fellow at the Saïd Business School at the University of Oxford, where she teaches in the Oxford Scenarios Program. She is the author of *Volatile Visions: Transactions in Anticipatory Knowledge* (Samfundslitteratur, 2006) and the coeditor of *Presenting Futures: Yearbook of Nanotechnology in Society* (Springer, 2008). She earned a PhD in Knowledge and Management at the Copenhagen Business School and a master's degree in Science and Technology Studies at Roskilde University in Demark.

STORMY
MUCH RAIN
RAIN
CHANGE
FAIR
SET FAIR
VERY DRY

Paul Rosero Contreras, *Arriba!*, Paradise Bay, Antarctic Peninsula, March 2017. Courtesy of Paul Rosero Contreras.

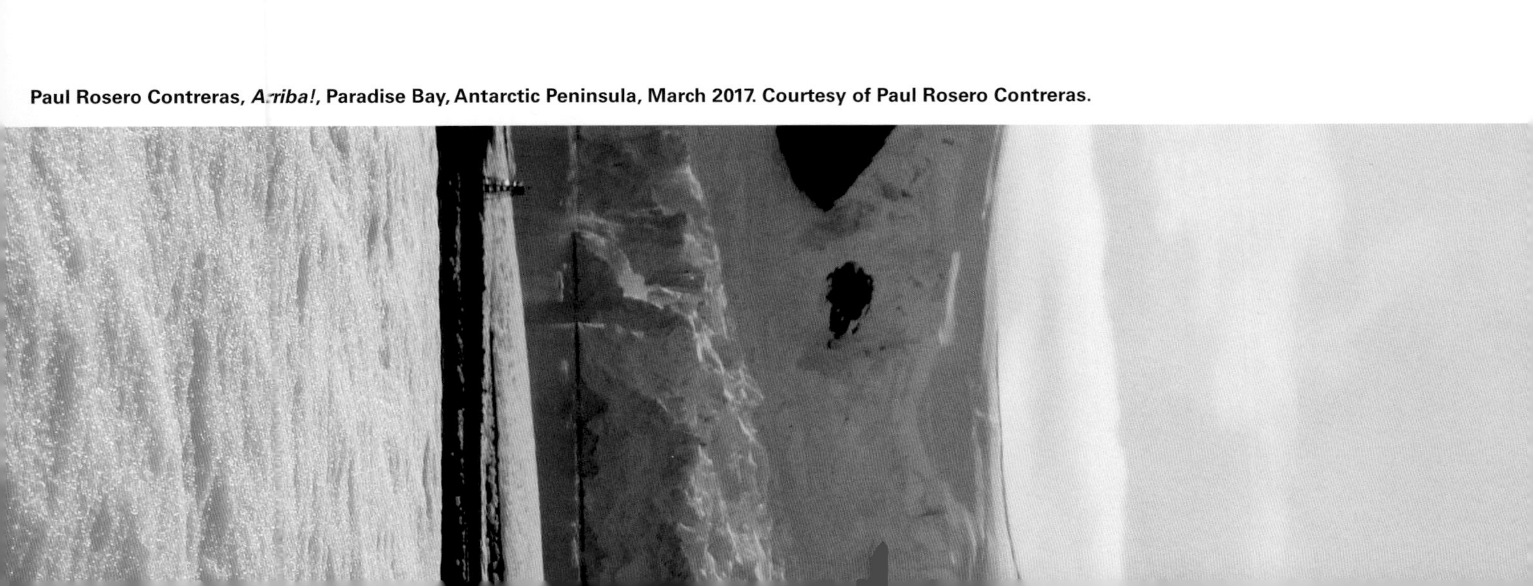

## Errant Curating

Nadim Samman

"And now, with the world before me, whither should I bend my steps?"

—Mary Shelley
*Frankenstein: or, The Modern Prometheus*

Antarctica (-64.8499966 -62.8999964): Piercing light across a bay of icebergs, whipped into shape by wind and sea, some curved, others striated, bearing stalagmites and deep blue fissures. A rocky shoreline, soon becoming steep, upward on one side approaching a lookout. On the far side of this ridge, a channel ringed by hulking glacial cliffs that fracture like carrara marble at the water's edge. Above them, rising into the clouds and out of sight, ice upon ice, eons of snow. Here, today—taking up less than 10 square feet of the continent's 150 billion—is a live cocoa tree that has traveled all the way from the Ecuadorian rain forest. Set within a life support system (a glass greenhouse whose interior climate reproduces equatorial conditions), its verdant leaves cast a surreal tint over the polar scene. This tree has wandered far from home. So has the artist who planted it.[1] Under the aegis of the 1st Antarctic Biennale, so has the institution of the biennale itself—and, most obviously, the curatorial mandate.[2]

---

1 The work, titled *Arriba!* (2017), is by Paul Rosero Contreras. It was conceived as a kind of tropical time capsule, referencing the fact that fifty million years ago Antarctica itself had a wholly different climate. Fossils of tropical flora have been discovered in the region where the work was installed.

**Andre Kuzkin, *99 Landscapes with a Tree*, Paradise Bay, Antarctic Peninsula, March 2017. Courtesy of the Antarctic Biennale.**

In light of the above, the following question presents itself: How has the site of the art exhibition wandered so far from its historical locus—the *home* of the muses?[3] Wasn't the curator's habitual task, until the 1960s, "hanging and placing" work on walls?[4] Today, the aesthetico-political shibboleth of a "big, beautiful wall" only bolsters Harald Szeemann's demand that we abandon this task.[5] Let us, therefore, offer some notes toward a theory of the curatorial agent as monster, pirate, authority, and

---

**2** The expedition of the 1st Antarctic Biennale, held under the patronage of UNESCO, left the shores of Tierra del Fuego, Argentina on March 16, 2017, and concluded with a ceremonial reception in honor of the biennale participants at the Faena Arts Center in Buenos Aires on March 29, 2017. Its commissioner was artist Alexander Ponomarev, and its co-curator was Nadim Samman. See http://www.antarcticpavilion.com.

**3** The museum.

**4** Harald Szeemann quoted in Carolee Thea, *Foci: Interviews with Ten International Curators* (New York: Apex Art Curatorial Program, 2001), 17.

**5** Tracy Jan, "Trump's Big, Beautiful Wall Will Require Him to Take Big Swaths of Other People's Land," *Washington Post*, March 21, 2017, https://www.washingtonpost.com/news/wonk/wp/2017/03/21/trumps-big-beautiful-wall-will-require-him-to-take-big-swaths-of-other-peoples-land.

pilot—by appeal to its environmental engagements. This essay is about wandering as curatorial method—about exhibition making in an errant mode, beyond galleries, indeed, beyond cities. Like Victor Frankenstein's monster, this curating traverses the planet, from domestic settings to geographical extremes—reconfiguring spatial, jurisdictional, identity, and narrative values through its (sometimes uninvited) presence(s). Just as the monster turned Victor's hearth into a site of horror, wholly upending familiar/familial relations, today's curatorial wandering rescores environments. While errant curating is, arguably, less overdetermined by tragedy, it is likewise a sovereign vector let loose into the world. It cannot be unborn.

### Curating Space—From Field to Vector

By the early 1970s, the domain of art had come to include "tons of earth excavated from the desert, stockades of logs surrounded by firepits," and more. Rosalind Krauss would argue for the novelty of this phenomenon, rebutting "historicist" attempts to establish a genealogy that would position such gestures as the latest in a lineage stretching back to prehistoric earthworks such as the Nazca Lines.[6] That perspective was, she asserted, blinkered by modernism's "definition of a given medium on the grounds of material, or...the perception of material." In contrast, Krauss staked out an analytic terrain whose predicates she arranged as coordinates, delimiting sculpture's "expanded field." This was an attempt to explicate a "totally new" program, to give certain artists their due as quasi-philosophers who "problematized oppositions" between concepts like "landscape" and (negations such as) "not-architecture."[7] As crystallized in the diagrammatic component of her argument, whose appearance elided hard-edged abstraction and hard theory, the heterogeneous agendas of individual

---

**6** According to Rosalind Krauss, some of the artists associated with this observation were Robert Morris, Robert Smithson, Michael Heizer, Richard Serra, Walter De Maria, Sol LeWitt, Bruce Nauman, and, later on, Robert Irwin, Alice Aycock, John Mason, Mary Miss, Charles Simonds, Dennis Oppenheim, Nancy Holt, and George Trakas.

**7** Rosalind Krauss, "Sculpture in the Expanded Field," *October* 8 (Spring 1979): 30–44.

makers cohered: it would seem that art was now capable of realizing all possible statements within a given language.

In expanding the theoretical parameters of sculpture to include negations (not-landscape, not-architecture), Krauss's linguistic turn pursued the postmodern project of ungrounding established cultural norms—an agenda corollary with social liberation struggles. However, historical distance affords us another perspective: one wherein the expanded field, in both practice and theory, symptomatized a broader, increasingly limitless appetite for control of space in public discourse. Cutting across the domains of high art, politics, and mass culture, a tide seemed to announce *It is ours, in any case*. Within this sovereign reverie, ground would not need to be marked by a permanent or obvious figure (such as an architectural one) in order to be occupied; claims could take place within not-landscapes and, indeed, might shape them as cultural domains—even when such not-landscapes resembled a void.

Contemporaneous with the artistic developments outlined in Krauss's article, state authorship was intensively reckoning with the abyss of outer space. Crudely relying on colonial patterns of old, the state began with a blunt desire to mark its non-site—Sputnik's ping acting as a sonic flag, preceding Yuri Gagarin's 1961 orbit, followed by the milestone planting of the Stars and Stripes on the Moon by the astronauts of Apollo 11 in 1969. Only four months later, the Apollo 12 mission carried a small ceramic tile purporting to be a "museum" to our nearest celestial neighbor. Inscribed on it were drawings by the American artists Robert Rauschenberg, Andy Warhol, Claes Oldenberg, John Chamberlain, Forrest Myers, and David Novros.[8] Given the all-male selection by artist-curator Forrest Myers, is it any wonder that Warhol's contribution was a penis? While these artists gained the outer reaches of spatial access, their names would not make it into Krauss's account of expanded practice. One surmises that their cultural touchstones were somewhat parochial—re-inscribing existing values onto a new site: Oldenberg's Mickey Mouse in

8 For the Moon Museum, as it was called, Rauschenberg drew a straight line; Warhol drew a penis; Oldenberg drew an image of Mickey Mouse; Chamberlain, Myers, and Novros all drew geometric designs. See Grace Glueck, "New York Sculptor Says Intrepid Put Art on the Moon," *New York Times*, November 22, 1969, 19.

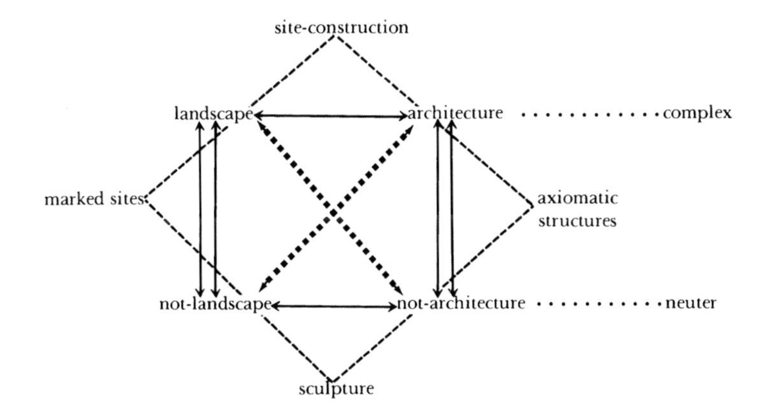


**Rosalind Krauss, diagram for the structural parameters of sculpture, architecture, and landscape architecture. Originally published in her piece "Sculpture in the Expanded Field,"** ***October*** **8 (Spring 1979): 38.**

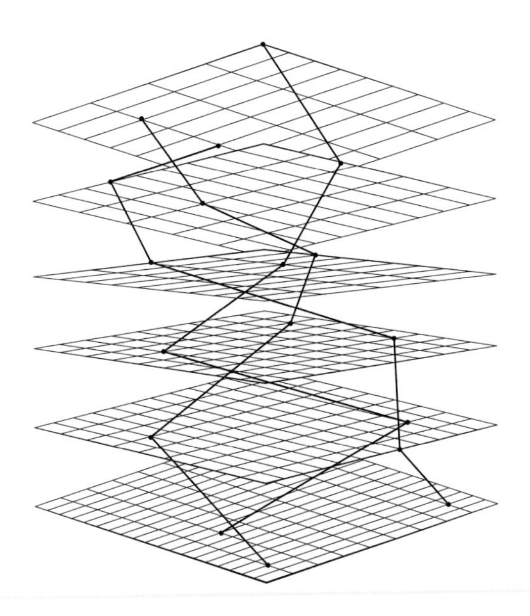

**Metahaven, diagram of The Stack, 2015. Illustrating Benjamin Bratton's concept of political geography as an "accidental megastructure" composed of six layers: Earth, Cloud, City, Address, Interface, and User. Courtesy of Metahaven.**

space too closely tracking the national political-symbolic agenda, despite the project's nongovernmental status.[9]

According to the curators of the 2013 *The Whole Earth: California and the Disappearance of the Outside*, the cultural values associated with 1960s space expansion that were truly new were not generated by looking out but by looking back—at our own planet. It was as a result of this endeavor that the photographic image of the terrestrial globe was born. Astronaut William Anders's *Earthrise* of 1968 would capture the entire earth in a picture. Later that year, the same image would appear on the cover of the first issue of *The Whole Earth Catalogue*. It would also grace Buckminster Fuller's *Operating Manual for Spaceship Earth*. Both publications were signal moments in popularizing the holistic "systems" thinking associated with the emergent complex of cybernetics, globalization, and environmentalism. While imagining the earth as a total mechanism, with mankind sitting at the controls, such publications exemplified an ascendant discourse of planetary stewardship—one wherein the avant-garde figure of the "artist-engineer" first proposed by Aleksandr Rodchenko might, we observe, following its Stalinist travesty, be rehabilitated. Between sculpture's expanded field and Spaceship Earth, total spatial authorship appeared to be a new given.

In practice, however, this totality was circumscribed. While the ideal plane of Krauss's diagram could deliver a universal field, and frictionless conceptual vectors traversing it, crossing (extra) terrestrial space is a matter of material exigencies, subject to the politics of access and exclusion. As the necessary smuggling of the Moon Museum indicated—circumventing NASA's lack of interest in facilitating the project—the new (non)sites of curatorial authority were not only novel coordinates (or nodes) but *vectors themselves*. Henceforth, apart from the self-aggrandizing challenge of accessing material networks in pursuit of expansive authorial distribution in space (that is, flag planting), a twofold task presented itself: making infrastructure (more) visible to critique and mapping/enacting its otherwise unidentified cultural affordances.

---

9 The Moon Museum was smuggled onboard by a sympathetic engineer after unsuccessful attempts by artist-curator Forrest Myers to gain approval from NASA. See "Who Is John F.?," *History Detectives*, PBS, season 8, episode 1, June 7, 2010.

According to a prominent member of the strategic studies community, contemporary society is undergoing "a fundamental transformation by which functional infrastructure tells us more about how the world works than physical borders."[10] And yet, this infrastructure also clouds our vision. It is disorienting, insofar as it advances epistemological equivalence, apparently disclosing a lack of privileged vantage points from which to survey an intellectual/cultural scene. Under this hyperlinked condition, one site gives way to another—both physically and conceptually. But at least part of the "post-truth" lie—and the seizure of sovereignty that it enacts—stems from the less than visible characteristics of infrastructure itself. As the task of disclosure grows more urgent, its mines, cables, server farms, and security apparatuses are beginning to feature in exhibitions.[11] Such projects follow the iconoclastic tradition of Hans Haacke, whose attention to the economic framework of one particular museum would martyr the muses, skewering their moral domicile by tracing financial flows.[12] But while they may have been turned to stone by the inscription of the infrastructure gorgon within their home, the latter was still very much alive. Its power not so much contained as spotlighted in order to promote public literacy. The expanded sovereignty of today's exhibition, however, must obtain the vector itself. That is to say, "working the world" consists in the *functional*: operating at least some of the control protocols of *spaceship infrastructure*.

My use of the term *spaceship infrastructure* proceeds from a reflection on the emerging wave of geopolitical theory, which sets out to "map" the technical integration of the globe in order

10 Parag Khanna, *Connectography: Mapping the Future of Global Civilization* (New York: Random House, 2016), xvi–xvii. He continues: "The true map of the world should feature not just states but megacities, highways, pipelines, Internet cables, and other symbols of our emerging global network civilization."

11 From an attention to the geology of media in the art of Revital Cohen and Tuur Van Balen to the mapping of transatlantic cables in the productions of Trevor Paglen and Lance Wakeling; from the labor conditions at Google scanning facilities, in a project by Andrew Norman Wilson, to Simon Denny's physicalization of the PowerPoint ideology of the PRISM surveillance enterprise.

12 See Hans Haacke's exhibition *Shapolsky et al: Manhattan Real Estate Holdings, a Real Time Social System as of May 1, 1971*, Solomon R. Guggenheim Museum, New York, NY, 1971.

to comment on the contemporary condition(s) of sovereignty. In the detail and methodological basis of their diagnoses, writings by figures from opposing ends of the ideological spectrum (such as Parag Khanna and Benjamin H. Bratton) dovetail in their invitation to consider Fuller's book less a manual and more a prolegomena to future ones. Keeping the latter's Spaceship Earth dyad in the frame, we may suggest that both thinkers' reflections are turned toward the spaceship. The ship, as anyone who has spent an extended period of time at sea is likely to reflect, is a totally designed environment—one whose architecture scores, in the manner of choreography, life onboard.[13] Herewith, Khanna's explication of living within: "There is no undesignated space [even] the skies are cluttered with airplanes, satellites, and increasingly drones, layered with CO2 emissions and pollution, and permeated by radar and telecommunications."[14] Bratton similarly speaks from the interior. His model "does not put technology 'inside' a 'society,' but sees a technological totality as the armature of the social itself." We "dwell within" an "accidental megastructure...a new architecture" that "divide[s] up the world into sovereign spaces."[15] This megastructure incorporates "infrastructure at the continental level, pervasive computing at the urban scale, and ambient interfaces at the perceptual scale," among other things.[16] For Bratton (though reflected in Khanna's comments on the sky), maps of horizontal space (planar geography) "can't account for all the overlapping layers that create a thickened vertical jurisdictional complexity."[17] There is, we know, more than one deck on a ship.

Authorship is, taking these considerations into account, a vector through various layers of jurisdiction—engaging each of their respective designations (or score) and (re)interpreting them in turn. What Bratton terms the "design horizons" of each layer must be probed for unintended affordances—pieces of open

---

**13** The work of artist-architect Alex Schweder, whose practice is self-defined as "performance architecture," is preeminently concerned with this condition.

**14** Khanna, *Connectography*, 15.

**15** Benjamin H. Bratton, *The Stack: On Software and Sovereignty* (Cambridge, MA: MIT Press, 2016), xvii.

**16** Bratton, *The Stack*, 3–4.

**17** Bratton, *The Stack*, 4.

source code; latent architectural possibilities. In some cases, layers may be totally redesigned. Beyond a purely material frame, we observe that "jurisdictional" accommodates the soft specificities of site identified by Miwon Kwon, namely "cultural debates, a theoretical concept...a historical condition, even particular formations of desire."[18] In light of all this, today's wandering curatorial enterprise—which we may term a *total* species of exhibition design—incorporates acts of renovation, rescoring/(re)interpretation, renegotiation, and even redesiring, in its (sovereign) operations.

Rather than the arrangement of objects on the walls of a gallery or museum, the first task of layered curating concerns the selection of jurisdictions that are to be put on display in a given project. The subsequent task of (critical) exhibition design consists in staging their relation to one another (drawing out their points of connection) as a system, and altering arrangements within each—arguing the case. In this process of rearranging, we have recourse to existing choreographies supplied by artists and the option of commissioning new works/levers. But we may not always need to deploy "art" in our operation—or, at least, deploy it correctly.[19] Moreover, in pursuit of real disclosure, the "making visible" proper to the concept of the *exhibit*, we may find ourselves wandering beyond museums, galleries, and perhaps cities. We may even leave, at least to a cursory view, exhibition audiences behind. But, in fact, it is in this errant mode that the author/ship of curating actually steers toward them.

In wandering, our procedure consists in toggling between (geographic) distance and functional proximity—an exemplary making visible of the latter. This mode of display (re)maps the audience's position in relation to "distant" locales, for the purpose of tracing the former's impact on them (ecologically, for

---

**18** Miwon Kwon, "One Place After Another: Notes on Site Specificity," *October* 80 (Spring 1997): 93.

**19** *Treasure of Lima: A Buried Exhibition*, which will be discussed in the next section, is a signal example of deploying art "incorrectly"—if the correct thing to do is not bury the artist's agency and artwork beneath a total(izing) curatorial proposition. *Treasure of Lima: A Buried Exhibition*, Thyssen-Bornemisza Art Contemporary Academy (TBA21–Academy), Isla del Coco, Costa Rica, 2014, https://www.tba21.org/#item--treasure_of_lima--567.

instance). Conversely, it illuminates the sway such spaces hold over one's immediate situation. This curatorial vector exhibits the condition of operative inseparability, binding seemingly heterogeneous economic, social, ecological, identitarian, and geographic domains. Rather than focus exclusively on "states and their divisions" (the demarcation of borders, which, we maintain, need not be interpreted solely in a geopolitical sense but which may circumscribe multifarious jurisdictions), we aim to deliver the exhibition as complex para-state and/or parasite. Rather than instituting visual and informational opacity, and spatial inaccessibility, within our projects in order to disenfranchise the audience, we do so in order to lay bare otherwise obscure mandates for authorship. We leave in order to return (the audience's space to them); we take in order to give.[20] And so off we wander...

## Vector (I): A Buried Exhibition

Three hundred miles hard sail from the sweltering Costa Rican port of Golfito, Isla del Coco juts up from the Pacific deep. There is no way to fly in, and you cannot visit without a government permit. Covering just nine of the ocean's sixty-four million square miles, its shallows are an important spawning ground for coral—a cornucopia of fish species and legions of predators. At any given moment, the island is circled by hundreds-strong schools of hammerheads and a resident population of sharks: some Galapagos, Silkies, Black and White Tips, and Sea Tigers. Above the water line, cliffs—broken only by waterfalls—support a seething, virgin, jungle.

Perhaps the only place in the world where treasure hunting is explicitly illegal, stories of what is buried deep within Coco have developed into myth, inspiring novels and genre fantasies for more than a century. The most famous myth concerns the so-called Treasure of Lima. In 1820, with the army of José de San

20 Sometimes we display complete author-ity over objects (such as artworks), subjecting them to iconoclastic conceptual appropriation. Done without a critical conception of (dis)play, this move is rightly abjured. However, when we are endeavoring to disclose how certain spaces, and the objects within them, are already appropriated, travesty is unavoidable.

Martín approaching, Peru's viceroy José de la Serna entrusted the treasury of the city's cathedral to a British sea trader named Captain William Thompson. Instead of remaining in the harbor as instructed, Thompson and his men slit the throats of the viceroy's men and sailed to Coco. They were later apprehended, and all of the crew except for the captain and his first mate were hanged. The lucky two were spared only after promising to guide their captors to the rich hoard.[21] After arriving on Coco, the pair ran off into the trees, never to be found. Several treasure hunting expeditions would later be mounted on the basis of claims by a man named Keating, who was said to have befriended Thompson on his deathbed in Newfoundland and received a map to the hoard. Before writing *Treasure Island*, Robert Louis Stevenson read about one such (failed) expedition in a San Francisco newspaper, and some have argued that the map he drew of his "fictional" isle closely resembles numerous treasure charts of Coco.

The 2014 exhibition *Treasure of Lima: A Buried Exhibition* involved burying an ensemble of commissioned works by thirty-nine prominent artists at a secret location on Coco, with the permission of the Costa Rican National Park Service. The nature of the works themselves would be kept secret. They were housed in a bespoke capsule designed by architects Aranda\Lasch, which looked nothing like a wooden treasure chest of old.[22] Polished stainless steel, the chest was a truncated tetrahedron that opened (like an oyster) to reveal a sphere at the center. This oversize "pearl" comprised a vacuum-sealed glass vessel, normally used to protect underwater cameras, housing a series of aluminum boxes that contained works on paper, small sculptures, LPs, and

---

**21** The original treasure consisted of precious metals, stones, and artifacts requisitioned by the Spanish from their Central and South American dominions, including 113 gold religious statues, one of which was a life-size Virgin Mary; 200 chests of jewels; 273 swords with jeweled hilts; 1,000 diamonds; numerous solid-gold crowns; 150 chalices; and hundreds of gold and silver bars. Its value today is reckoned around $200 million. See Jasper Copping, "British Explorer Closes in on Legendary 'Treasure of Lima,'" *Business Insider*, August 5, 2012, http://www.businessinsider.com/british-explorer-closes-in-on-legendary-treasure-of-lima-2012-8.

**22** Rather than conforming to an archaic cliché, the chest recalled the product design of market-leading personal computing hardware. While by no means illustrative, its hygienic surfaces and acute angles intentionally suggested a kind of oversize digital-data-storage device.

video and sound files stored on a hard drive. Upon its arrival at Coco, the chest was brought to shore on a raft, after which it was man-hauled inland, where it was eventually interred at a suitable location. The GPS coordinates of the site were logged, and upon returning to Berlin, the only extant copy of the coordinates was turned over to the Dutch artist Constant Dullaart. Working with a leading IT security consultant, who remained anonymous, and following strict data-protection protocols, Dullaart oversaw the coordinates' encryption to the highest possible specification. The resulting cypher, over two thousand characters long, was then given physical form—3-D-printed into a steel cylinder.[23] The cylinder was placed in a second, unburied, version of the chest, which was auctioned off at a New York evening sale. Purchased for $185,000, the anonymous buyer possessed an all but "unreadable" map (no de-encryption key was supplied), to a collection or "exhibition" of works hidden on an island unreachable without a permit and where digging for treasure is explicitly illegal. In collaboration with Costa Rican partners, the Thyssen-Bornemisza Art Contemporary Academy (TBA21–Academy), which commissioned the project, used this money to initiate a research and conservation project for the sharks inhabiting the island's surrounding waters.[24]

---

**23** Constant Dullaart's *Map* (2014) is both a sculpture—a unique physical object whose form was determined by an artist—and a tool or set of instructions for disclosing an elsewhere. On the one hand, its cylindrical form serves to recall antique maps or scrolls (an explicit reference to Coco's maritime history), staging its unreadable script as a digital-era successor to the idiosyncratic markings inscribed on the pirate charts of legend. On the other hand, it is also a feature of the encryption system: without any indication of where the code begins or ends, the code is exponentially harder to crack. Yet this design as resistance is contradicted by *Map*'s utility for the would-be code breaker, which allows the sculpture to be used as a rolling printing plate—enabling the physical transfer of data to paper by way of ink. With *Map* our project dramatizes the interconnection between the physical and the informatic. These considerations raise the following questions: Must *Map* be used, rather than contemplated, in order for it to achieve the status of an artwork? Or, rather, does it only remain an artwork if its functional indeterminacy is maintained?

**24** TBA21–Academy and Alligator Head Foundation collaborated with Parc Nacional Isla del Coco and Costa Rican partners La Fundación Amigos de la Isla del Coco, and Misión Tiburón. Tagging sharks at Isla del Coco began on September 6, 2018, and data collection by satellite telemetry revealed migratory routes as far east as the Columbian coast. As well as illuminating the life of these little studied animals, hard data may now be used to make the case for regional cooperation on conservation initiatives.

Aranda\Lasch, *Treasure Chest, Treasure of Lima: A Buried Exhibition*, TBA21–Academy, Isla del Coco, Costa Rica, 2014. Photograph by Alex Gruber. Courtesy of TBA21–Academy.

*Treasure of Lima: A Buried Exhibition*, TBA21–Academy, Isla del Coco, Costa Rica, 2014. Photograph by Julian Charrière. Courtesy of TBA21–Academy.

More than a physical incursion, *Treasure of Lima: A Buried Exhibition* effected a rescoring/rearrangement of the narrative/historic, legal, economic, and biological coordinates of Coco—one that, simultaneously, exhibited this complex set of relations. Clearly, the project's title was borrowed from the original Treasure of Lima. This doubling was calculated to produce a misfiling within the historical archive, whereby—in addition to being the first buried treasure on the island in two centuries—the smuggled content would expand the definition of "treasure" in the course of future research into Coco's history. In particular, this move was ventured to impact narratives concerning ownership, exploitation, misappropriation, and colonialism. Hijacking the maritime dimensions of Central American history (in its pirate element), the project set these topics into relief, dramatizing modes of value and methods of identification in the present and speaking to the topic of nonfinancial (biological) worth.[25]

The project further strove to exhibit the regulatory and enforcement framework by which Coco's natural "treasures" are secured, on ecological grounds; also, to rescore interest in this system through the act of display. By negotiating permission to bury our exhibition—what real pirate would do that?—we acknowledged the existing regulatory/jurisdictional safeguards. However, by design, the endeavor instituted a possible threat to their enforcement: the intervention, publicized worldwide, certainly aroused interest in the chest's potential recovery. If anyone, at some point in the future, gains access to the works contained within it, it will indicate a failure to enforce protection statutes or their abolition entirely.[26] Curating as inoculation.

In economic terms, the project financially resourced a shark research and conservation scheme. This new project, itself, aimed to produce a resource of scientific information that would inform future biological management and regulation. Convening

**25** The island was declared a national park in 1978 and UNESCO named it a World Heritage Site in 1997. The Seamounts Marine Management Area—the aquatic reserve created in 2011 that surrounds the island—is larger than Yellowstone National Park and second only to the Galápagos National Park in terms of protected marine areas in the Tropical Eastern Pacific.

**26** Under such circumstances, the recovery of the buried exhibition will mark an assault on something of greater value.

***Treasure of Lima: A Buried Exhibition*, TBA21–Academy, Isla del Coco, Costa Rica, 2014. Photograph by Julian Charrière. Courtesy of TBA21–Academy.**

***Treasure of Lima: A Buried Exhibition,*** **TBA21–Academy, Isla del Coco, Costa Rica, 2014. Photograph by Julian Charrière. Courtesy of TBA21–Academy.**

experts to draft an outline for this scheme was, it should be noted, a component of the curatorial enterprise.[27] In light of rampant poaching in the national park's waters, the "exhibition" sought to diagnose the impact of ecological piracy and, additionally, of palliative agendas—directly approaching the "cure" promised by the word "curatorial."

Effecting a layered/vertical dramaturgy, the various "burials" involved in the project manifested a translocational situation of authorship. Beginning with the statement of buried artworks (announced as a practical fact involving mud and shovels), the

---

**27** As a communication and consultation exercise to develop the "Pelagic Research and Conservation Project for Isla del Coco," the TBA21-Academy convened a symposium at UCLA's Ideas Campus on August 11, 2014. The symposium also highlighted shark conservation initiatives in the Tropical Eastern Pacific and was accompanied by a dive expedition around Catalina Island. For more information, see http://www.tba21.org/#item--los_angeles--580.

*Treasure of Lima: A Buried Exhibition*, TBA21–Academy, Isla del Coco, Costa Rica, 2014. Photograph by Julian Charrière. Courtesy of TBA21–Academy.

project further entombed the GPS coordinates of the chest within the virtual crypt(ography) of Constant Dullaart's *Map*. While denying the gathered artworks a chance to speak for themselves, burying them beneath visual and narrative mediations, the exhibition's "making visible" abandoned the apparent antecedence of art documentation.[28] Considering the symbiotic relationship between *Map* and *Treasure Chest*, in terms of both symbolism and functionality, as well as their relation to the thirty-eight artworks contained within them, the venue for *Treasure of Lima: A Buried Exhibition*'s was not identified with any one layer-site but, rather, an entire complex spanning geographical distances and functional proximities.[29]

Extending into the space of the art market, inscribing the actions of the collector within the exhibitionary proposition, the project *contains* an implicit question regarding the exchange value of the works deployed in the project: whoever bought *Map* might have a better chance of recovering the buried *Treasure Chest*, but this neither guarantees practical nor legal possession. First, the buyer purchased *Map* without a key to unlock its sophisticated encryption. But there is also the issue of gaining access to the island, and given that digging for treasure is banned, this might involve breaking the law. Purchasing the map does not necessarily underwrite ownership of the buried artworks, even if they are eventually recovered. Legally speaking, only the physical *Map* was sold at auction. In this respect, the potential ownership of the buried artworks is, itself, buried beneath a set of challenges. This leaves the buyer with a performative score that they are free to interpret—or rather, display/make visible within the field of the exhibition. They will have acquired some

---

**28** The project took inspiration from Boris Groys's insight that, today, art documentation enjoys a status that approaches near equivalence with the artwork itself; and a recent assessment that broadcast media practice was a vital dimension of historic Land Art. The project attempted to exhibit this condition and explore its potential.

**29** These comments shed some light on how surveillance—observation—relates to our new figure. Following Edward Snowden, it would seem that enclosure within a cryptographic strongbox allows for paradoxical identification: to be one thing and another simultaneously. On a political level, personal data protection helps us to maintain a translocational identity that amounts to freedom itself. When we are observed and measured as one thing or another by an external gaze our paradoxical potential—to be both outlaws and good citizens, for instance—is dead in the water. We are collected, put on file—some butterflies pinned, others broken on a wheel.

of the means by which to recover an amazing art collection, but what matters to them most? They will either put *Map* to use in an attempt to recover the works buried on Coco or enact a relation to both the superposition of art in the project and the value of restricting human access to the island. The latter is a relation of trust.[30] Herein, a (re)scoring of desire: ascertaining the GPS coordinates might allow one to drop a pin on a map of Coco, to thrust a spade into the soil, and to ultimately observe the contents of the chest. But in this process something will be lost: once opened, *Map* is just a map; the exhibition, just a set of things in a box buried on an island. Once opened, the curatorial operation is complete/over.

The aforementioned trust is of a piece with a general outlook that recognizes no operative separation between nature, culture, and humanity. What appeared to be an island separated from other lands is just one part of a larger system. Rather than there being a yawning gap between the sharks of Coco and metropolitan modernity, there is only interconnection and engagement. We cannot avoid impacting these creatures, either by focused exploitation or laissez-faire fallout. As the curatorial rubric of the project makes plain—we must, today, critically appraise the design of relationships that span oceans and continents. We must curate them.

### Vector (II): A Wandering Biennale

In March 2017 an international group of sixty-five artists, scientists, architects, and philosophers left the port of Ushuaia, Argentina—bound for the Antarctic Circle on board the *Akademik Sergei Vavilov* (part of the Russian Shirshov Institute of Oceanology's scientific research fleet). The voyage covered approximately two thousand nautical miles and made landfall at twelve sites in the Antarctic archipelago over a period of two

---

**30** Trust relinquishes control of immediate benefits—in this case, the collector considers proximity to the buried art objects or the island itself phenomena whose quantification does not necessarily indicate the quality of the relationship: they trust that geographic distance, and seeming lack of access to the art objects, is made up for in functional proximity by moral/ecological quality.

weeks before returning via Cape Horn.[31] At each location, installations, sculptures, exhibitions, and performances were realized. Mobility, site specificity, and ecological compatibility were key touchstones. Nothing was left behind and no audience was present—notwithstanding the participants themselves and Antarctica's native species. Actions included a landscape photography exhibition for penguins (they didn't seem to get much out of it) and an underwater installation for whales.[32] In addition to land and sea interventions, the *Vavilov* served as a floating studio, photo lab, exhibition, performance, and conference facility. Onboard activities included fifteen symposia, incorporating alternative histories of south polar enterprise, and a daily screening program featuring commissioned videos.[33] Throughout, our discussions focused on future cross-disciplinary collaborations and on the question, "What potential does the Antarctic imaginary hold?"[34]

---

**31** Anchors dropped in Neko Harbor, Paradise Bay, and Orne Harbor; Cuverville Island, the Errera Channel, the Lemaire Channel, Pléneau Island, Petermann Island, the Penola Strait, Deception Island, and elsewhere.

**32** In total, over twenty artistic projects were carried out, including performances, installations, exhibitions, and sound-art experiments by artists: Alexis Anastasiou (BR); Yto Barrada (MR); Julius von Bismarck (DE); Julian Charrière (FR/CH); Paul Rosero Contreras (EC); Gustav Duesing (DE); Zhang Enli (CN); Joaquín Fargas (AR); Sho Hasegawa (JP); Yasuaki Igarashi (JP); Katya Kovaleva (RU); Andre Kuzkin (RU); Juliana Cerqueira Leite (US/BR); Alexander Ponomarev (RU); Shama Rahman (UK); Abdullah Al Saadi (UAE); and Lou Sheppard (CA). A manifestation of the Aerocene project by Tomás Saraceno (AR) also took place. For further information about onboard projects, including video program, see http://antarcticbiennale.art/wp/category/expedition.

**33** The symposia series, convened by Nadim Samman and coordinated by Sofia Gavrilova, was titled "Antarctic Biennale Vision Club." This was the main platform for interdisciplinary participants: Elizabeth Barry (USA); Adrian Dannatt (UK); Barbara Imhof (AT); Wakana Kono (JP); Carlo Rizzo (IT); Alexander Sekatskii (RU), Jean de Pomereu (FR); Susmita Mohanty (IN); Hector Monsalve (AR); Miguel Petchkovsky (AG); Sergey Pisarev (RU); Nicholas Shapiro (US); Lisen Schultz (SE). Video series included work by Adrian Balseca (EC); Yto Barrada (MR); Emmy Skensved and Gregoire Blunt (CA); Julian Charrière (FR/CH); Paul Rosero Contreras (EC); Marcel Dinahet (FR); Constant Dullaart (NE); Karin Ferrari (AT); Etienne de France (FR); Swetlana Heger (SE); Young Hae-Chang Heavy Industries (KR); Eli Maria Lundgaard (NO), Eva and Franco Mattes (IT); and Jessica Sarah Rinland (UK).

**34** Three resulting collaborations are discussed in a paper coauthored by the editor of this volume, delivered at the 2017 International Astronautical Congress. All involve artists' engagements with the Self-Deployable Habitat for Extreme Environments (SHEE) prototype developed by Liquifer Systems Group for the European Space Agency. See Dehlia Hannah and Barbara Imhof, "Art in Extreme Environments: Reflections of Space Research from the 1st Antarctic Biennale," proceedings for the 68th International Astronautical Congress (IAC), Adelaide, Australia, September 25–29, 2017.

According to the 1959 Antarctic Treaty, the southern continent is reserved for peaceful scientific inquiry. Owned by no individual or nation (with sovereign claims suspended), Antarctica's legal-institutional framework is arguably the most successful example of international cooperation in modern history. Given that the system emerged at the height of the Cold War, this fact is doubly impressive. Moreover, it was initiated by a group of scientists rather than politicians. The exclusive right that the treaty accords scientific enterprise, incorporating a proscription against the exploitation of natural resources, is justly celebrated as a model for global conservation initiatives. However, in a deeper sense, the treaty can be viewed as much more than this: as a foundational document for *a new form of universal community*. Indeed, the treaty system suggests an incipient supranational identity based on cooperation and a sophisticated regard for ecology whose relevance transcends whatever takes place on the continent. It was this implication that we—Alexander Ponomarev and myself, the founders of the 1st Antarctic Biennale—believed must be made more proximate, for those persons who do not work with (or in) the region.

Even before the commissioning of individual artworks, the establishment of the Antarctic Biennale was ventured as an assertion of Antarctica's status as a potent cultural paradigm. Our operation proceeded according to the idea that Antarctica affords an imaginary that is most prescient yet underexplored: following the USSR's collapse, the figure of the "new Soviet man" was consigned to the scrapheap. Almost immediately, the American political philosopher Francis Fukuyama proposed liberal democracy as the "final model" for the "coherent and directional" development of civil organization; the history (of competing formats) had ended, he argued, and the liberal-democratic subject was the "last man" standing.[35] Today, this is a wholly discredited idea. Foregrounding the (otherwise) repressed cultural dimensions of Antarctica through the proposition of a festival, our action issued from the hypothesis that the "Antarctic man" is a plausible world-historical successor. Beyond abstract formulation, this political

**35** Francis Fukuyama, *The End of History and the Last Man* (London: Penguin, 1992), xxii.

subject outstrips (in legal principle) the paradigm of the nation state. This subject also incorporates a holistic view of the planet as a complex—unified—system. Furthermore, with regard to an anthropological frame, it is a worthy hypothesis: more than a century after man (men) first set foot there, Antarctica sustains a population of 1,162 throughout the sunless winter and 4,000 in the summer months. Given the time that has elapsed and the amount of human activity, why not speak of culture (beyond official mission structures) peculiar to Antarctica? Perhaps, therefore, we should view it as the last terrestrial *new world*. Might not activities in Antarctica amount to a whole new set of customs, architectures, and attitudes of relevance beyond the bases? Exploring this idea, our biennale was ventured as a (post)nation-building initiative—a manifestly cultural festival of and in Antarctica.

Proceeding according to this outlook, the biennale's pre-expedition communications maintained that Antarctica is underexploited, not in a physical sense but as field of visual and conceptual inquiry. The Antarctic imaginary belongs to everyone, and yet, we claimed (on public stages in Italy, the United States, Sweden, Spain, Argentina, and Russia) control over the regime of images issued from this region is centralized.[36] For the most part, mimetic production is supplied by documentary photographers and filmmakers "embedded" within national-scientific brigades or else adhering to hegemonic interpretive frames. Thus, what passes for Antarctic "cultural" activity assumes a subordinate role to the "useful" research being carried out on bases, or the keynote messages issuing from them—via official media relations. Within this structural condition, there is little place for non-debentured, heterogeneous, representations of the Antarctic imaginary—no bottom up.[37] And we are all too aware of the limits of embedded reportage...

---

**36** At venues including the Ca' Foscari University, Venice, 2014; the Royal Geographical Society's conference "Explore," London, 2016; the Explorer's Club, New York, 2016; the Grenna Museum, Sweden, 2016; the Maritime Museum, Barcelona 2016; the New York Yacht Club, New York, 2016; and the Pushkin Museum, Moscow, 2016.

**37** Communications around the exhibition *Antarctopia*, at the inaugural Antarctic Pavilion for the 15th Venice Biennale of Architecture in 2014, argued that a lack of non-debentured reflection on Antarctic enterprise facilitates a geopolitical scramble for resources. *Antarctopia* was curated by myself and commissioned by Alexander Ponomarev. See http://www.antarcticpavilion.com/antarctopia-essays.html.

Under these conditions, the contours of the Antarctic man lack clarity and are only being discovered in a haphazard manner. If we are to realize Antarctica's potential for *those who cannot go,* then cultural workers must seize the means of south polar (image) production. In our view, it is only through intensified (and truly independent) engagement with Antarctica that we may discover—through aesthetic experimentation—its otherwise inaccessible intellectual, social, and political topography. It is this landscape that offers the most promising ground for harvesting radical images. Given the lack of an extant academic project addressing Antarctic art history, the creation of a biennale assumes the character of a demand that this be undertaken.[38]

In addition to the inspiration supplied by Antarctica's supranational administration, we deemed the creation of an independent platform from which to engage the continent's environment and science as timely. We will not see ourselves as *one* until we can view the biosphere, which encompasses our civilization as much as it does icebergs, as an integrated unit. Statistical proof of climate change and picturesque photographs of glaciers and penguins can only be so effective in altering the global public's self-image. The world requires a new regime of interdisciplinary image making *from below*; the overdetermined outcomes of government-run residencies are not enough. Bringing together artists, scientists, and thinkers, the 1st Antarctic Biennale was established not just to cultivate artists' engagements with a space reserved for "science" but to widen the scope of what is considered cultural while subjecting techno-scientific (post)sovereignty

---

**38** Antarctica is not generally communicated as a cultural space, despite the fact that it has been inhabited for more than a century. The lack of will to cultural analysis is clearly indicated by the dearth of comparative studies on its artistic production. Indeed, there have been no attempts to investigate *image-making* practices in Antarctica across expeditions or personalities. While attention has been paid to the work of creators individually, no definitive "art history of Antarctica" has been written. Even excepting the will to intellectual synthesis, libraries are without anthologies of paintings or drawings made there. Given the opportunity to successfully catalog, without omission, the entire corpus of such images from 1840 until the late twentieth century, it is surprising that no one has tried. The same can be applied to the question of Antarctic architectural history—a question asserted at the Antarctic Pavilion's *Antarctopia*. For a full statement of this argument see my own piece, "The Antarctic Imaginary: Pavilion for the South Pole," *uncube*, June 20, 2014, http://www.uncubemagazine.com/blog/13521899.

to a gentle challenge. This is to say, the biennale was as much about making visible (exhibiting) an image of a new cultural institution as a series of art objects.

With a limited on-site audience—indeed, with only participants present—the biennale departed from standard models of perennial exhibition making and viewing. As such, it was the calculated performance of a leap beyond the luxury ghetto of what passes for the contemporary art *world*. Doesn't the very term ring sweaty when mentioned in the same breath as the Ross Ice Shelf? Something other than miles separates this paradigm from the polar one. We who walk the baking flagstones of San Marco, who press flesh at the prosecco intermezzo—where chatter flits from the surface of the artwork to the decoration of the palazzo and the cut of a dress—are very far removed from encounters with the non-man-made environment. In fact, the Venice Biennale—the world's oldest and most prestigious of art festival—takes place amid entirely constructed terrain, where there are more stone carvings of flowers than real ones. Against the pageant supplied by this model biennale, Antarctica is a place that does not forgive hubris easily, a place where people sometimes eat their boots to avoid starving (and where sometimes expecting ice is too much).[39] "This is a biennale 'upside down,'" said my collaborator. "Instead of the usual national pavilions—the icy inaccessibility of the Antarctic continent. Instead of pompous apartments and hotel rooms—ascetic cabins. Instead of chaotic wanderings, through receptions and tourist-filled streets—a dialogue with Big Nature and an explosion of consciousness facilitated through the dialogues with scientists, futurists, and technological visionaries." As we conceived it, the Antarctic Biennale platform floated an *image* of succession from the hothouse of subterranean commercial dealing, spectacle, and social climbing that envelops the art of our times. The question of whether one can, or must, live up to an image remains open, however.

The 1st Antarctic Biennale was, both literally and metaphorically, a vehicle for facilitating *independent* cultural production in the south polar region. It was a mechanism for expanding the

---

**39** In order to fully realize his project, the German architect Gustav Duesing required below-zero air temperature within a given time. We did not encounter this condition.

Antarctic imaginary through aesthetic exploration and *interdisciplinary* encounters pursuing culture in a sense not limited to art. The theme for the edition borrowed Captain Nemo's motto from *20,000 Leagues Under the Sea*: "Mobilis in Mobili," meaning "moving amidst mobility." Traversing the Southern Ocean, passing through the Bransfield Straight, between craggy peaks and glaciers, down through the Lemaire Channel, and into the Antarctic Circle, it was an expedition as festival. But the movements to which the motto referred also encompassed a trajectory through shifting currents in climate science, changes in ice-sheet cover, geophysical dynamism, and biological upheaval. Lastly, the title embraced a movement—or vector—cutting across developments within various disciplinary spheres.

Both the buried exhibition and the Antarctic Biennale exemplify curatorial operations as lines of flight, leaving behind museums and audiences in order to reframe them. As a case study in errant curatorial authority, *Treasure of Lima* indicates how a contemporary exhibition may be an *inter-site-specific* performance. This mode of action is a sovereign wandering through spaceship infrastructure—and a *wondering* with it, along a trajectory running through the layered condition(s) of placehood. We errant curators "leave" our walled cities (our green zones) for geographical extremes, not to escape but to make the contemporary hearth more visible—spaceship infrastructure is always already *everywhere*, as far as we are concerned. Just like Isla del Coco, Antarctica is far from accessible for most of us. And yet, this geographical end of the world figures centrally in global conversations concerning climate change, underpinning claims about the *end of the world as we know it*. As such, through this combination of physical remoteness *and* inscription within a shared imaginary, it encapsulates a key contemporary condition. If a curator wanders toward a place like Antarctica, it is because all putative elsewheres are already coming for us. Antarctica, like the rest of the biosphere, has been animated by technoscience and would now speak to us—of our errors and our doom. Embracing this conversation—the likes of which Victor Frankenstein refused—our exhibition making talks (as monster) to the monstrous, (as anthropos) to the anthropocene. We do so in order to find a way to live together, as there is nowhere else to run.

Thor Heyerdahl, *The Kon-Tiki Expedition: By Raft Across the South Seas*, trans. F.H. Lyon (London: George Allen & Unwin Ltd., 1948), 27. Photograph of the Kon-Tiki expedition copyright Everett Collection Historical/Alamy Stock Photo.

## Monsters of the Deep

Hilairy Hartnett in conversation with Dehlia Hannah

Hilairy Hartnett is an oceanographer, bio-geochemist, and a leader of PlanetWorks—an interdisciplinary initiative focusing on planetary management and geoengineering, based at Arizona State University. Hilairy shared her research with the science fiction authors who convened for the reenactment of the "dare" and spoke about the ways that climate change is affecting the oceans. Like the natural philosophical investigation of galvanism that inspired Mary Shelley during the summer of 1816, today's cutting-edge science is both horrific and fascinating.

HILAIRY HARTNETT I've read some of the stories you sent, and my first reaction is, these are really cool...and kind of *dark*.

DEHLIA HANNAH Well, *Frankenstein: or, The Modern Prometheus* is a paradigmatic horror story, and the reenactment of the dare could certainly be understood as a provocation to speculate about dystopian futures. These stories *are* dark, but like the novel, they're also filled with moments of light; well-marked alternatives. What you told us about climate change and our oceans—now that's a dark story. How do you face that bleak future?

I wonder if people's gut response to the unknown is that it's likely to be bad or frightening. We humans don't respond well to lack of information. It shows up in literature as horror. In science, it shows up as a deep-seated feeling that there's an information gap we need to fill. And what do we do? We go out and collect data. That way we can interpret it, describe how the system works, and decide what to think and what to do. Once you know how something works, it is suddenly much less scary. It may not be good, but at least you know what is happening. I think there's a lot of that fear of the unknown built into the climate story, especially in relation to the ocean.

Our aesthetic sense and understanding of the world has always had something to do with the ocean. It's in our mythologies, and it underpins our modern, global, industrial complex. Yet most people's experience of the ocean relates to the surface. We see the waves on the beach and experience them as active expressions. We kind of perceive the ocean as a living thing, because it moves; it flows. But there's more to it. As an oceanographer, it is clear just how huge the ocean is, that the vast majority of it goes unseen. At immense depths, the ocean is a very different place. It's dark and unknown, for the most part, and yet we know that this massive ocean has an incredible capacity to moderate our planet.

The oceans are like a thermostat. It's really hard to heat or cool water. That's why it takes time for a pot of water to boil or to make ice. The ocean *doesn't want* to change its temperature. You have to exert a *lot* of energy to make this happen—and we're doing it. You can see, now, that the oceans are warming. It has taken two hundred and something years of industrial revolution and global warming to do it, but it is happening.

It is as if the ocean has this giant, slow soul... its chemistry is working on a very long timescale. There are certainly things that ocean chemistry does on a timescale of seconds, but other processes take thousands of years. One of these processes is the way the ocean absorbs CO2. Through photosynthesis, algae take up CO2. The ocean is an enormous carbon sink, but there is a limit to how much the ocean can absorb. Slowly but surely, the CO2 in the atmosphere is increasing, and the ocean is kind of at capacity in terms of how much CO2 it can take up and release. Every day it does the same thing it has always done, but because of the long-term absorption of $CO_2$, we may be reaching a point where the ocean doesn't do this so well.

Is that something we can already see?

It is worth remembering that humans live at relatively small scales. We're about six feet tall, and weigh around two hundred pounds. We're not used to dealing with things at an immense scale, or at a tiny scale. Asking people to reflect on a planetary event or problem is hard, partly because humans don't process extreme timescales. We think that the ocean will be fine, in the long term, that the ocean is not worried—I don't know if the oceans worry, but I like to think they do—about the Anthropocene and human $CO_2$. Things will change, and things will change back, but over a ten-thousand-year process. It's like we're pushing on a giant, heavy rock, and it's not moving very much. But if it moves just a little, it's going to take a long time to get where it's going and a very long time for it to settle back to where it started. It's like poking a monster.

Monsters of the deep are an old idea. The ocean is dark and full of creatures, full of big monsters with sharp teeth. It is a place where new species are still being found all the time. But the idea that the ocean itself is a monster, which if disturbed or awoken may participate in the earth system in an altogether unwelcome manner—that's scary and fascinating.

I see it as fascinating. It's funny, I find the surface scarier. Tsunamis and big waves. That's all extremely frightening. Though, I've never been down in Alvin or to the bottom of the ocean. Maybe that's scary too—but I would love to do it, because I want to *know*. There are forests of tube worms around hydrothermal vents down there. They're not scary; they're amazing! There are all these beautiful phosphorescent and luminescent things, that, if we could see them, would be spectacularly gorgeous. The unknown is full of aesthetic beauty too.

Maybe this is where its gets philosophical: I'm trained as a chemist, and that means I have taught myself, over the years, to see molecules in space. I imagine molecules doing things that animals do. The chemistry of the ocean is beautiful. It is this relatively stable equilibrium-based system, where, if you push on it on one side, it responds—but eventually (on some timescale) it will slide or settle back. The thought that it has this almost homeostatic capacity to adjust its chemistry is amazing to me. It's hard to explain. You can't see it. You can't make a movie of ocean chemistry.

Oh right, dynamic equilibrium. How do you conceptualize constant change and stability at the same time? I love talking with scientists because you see things with your mind's eye that don't register without the relevant concept or measuring instrument. At the PlanetWorks conference we asked everyone: What timescales do you think in? You said seconds. Your fellow biogeochemist Arial Anbar said eons. I said hundreds of years—the timescale of the history of science and philosophy. Now that we're talking about climate change and knocking the ocean's pH out of balance, it seems like we're slipping between different timescales ...

When we start to erode that stability, the ocean story gets a little scary. The ocean's surface water exhibits a rapid response to CO2 in the atmosphere. You're familiar with the idea of CO2 in a liquid if you've ever had seltzer, soda, etc. That's CO2 dissolved in water, and it comes out in bubbles when the water is oversaturated. When CO2 in the atmosphere goes up, CO2 in the water goes up; and when CO2 in the atmosphere goes down, CO2 in the water goes down. We have pretty good data from the 1970s on seawater CO2, and it shows clearly that it's rising at a steady pace consistent with that of the atmosphere. CO2 is an acid, so when the ocean absorbs more than it can release, it starts to get more acidic. This has an impact on marine life, which will start to have an impact on us. I can't say, with my scientific hat on, exactly what is going to happen in the next 10, 50, 150 years, but I can tell you what the trend looks like, and that it is a little distressing. We're changing the ocean and not just by polluting it. The fundamental shifts in ocean chemistry are a really big deal. The ocean is going to take a long time to feel the full effect of the shift, and it's going to take an even longer time for the ocean to re-equilibrate and get back to whatever its new steady state is going to be.

This is where scale becomes important. One thing we forget is that the pH scale is logarithmic, so that a pH of 8 and a pH of 7 are a factor of 10 different. People think this is a small change because we count in ordinal ways on our fingers, but a change from pH 8 to pH 7 is a 10 times change, not a 10 percent change! If your salary changed by a factor of 10, you might be upset about it.

That's a helpful analogy. No matter how low or how high your salary was to begin with, you'd be accustomed to living within that allowance. Maybe you could get by with ten times less, and people do, but your life would be very different.

We've invented this weird system to talk about pH so that you can use ordinal numbers to talk about tiny changes in concentration. The Keeling Curve tells us that CO2 in the atmosphere rises and, as we now know, that the ocean acidifies at a predictable pace of 0.001 pH units per year. How many years until you have 0.01 or 0.1 or 1 pH unit? Not all of this can end up in the story, I'm sure...

Not all the numbers can end up in the story (and many of them didn't), but I find it fascinating to try to imagine the asynchrony between air time and ocean time. In literature, there are many ways of playing with our sense of time, by conveying the notion that the reader has a different vantage point and experience than the characters within a story, for example. It's amazing to think about the atmosphere changing so much faster than the ocean, and that some critters might notice that while other species—like us—take a whole lot longer to clue in about something changing.

Here is where there's a spectacular disconnect. The whole atmosphere mixes on a timescale of three years. On a statistical basis, if you take a molecule that was in Australia last week, I will breathe that molecule sometime in the next three years. We are completely connected at the surface, just like we are completely connected as a global population. But we are also completely connected in the deep ocean, on a very long timescale of about a thousand years. If we go into the earth's crust, the continents are also connected at a timescale that we don't have any perception of. We can barely perceive at a thousand-year timescale. If we're lucky, we can trace our ancestors back a thousand years. Our human experience doesn't give us good connections to timescales that are really long, and that's, I think, a big part of why climate change is problematic. We feel it on the short timescale. But we're willing to acknowledge that the surface is chaotic and noisy. We don't know if it's a real change or if it's just a couple years' change. When you start to see something that has been stable for a thousand years change, then you might wake up to the fact that maybe we're doing something. If we've done something to our planet that has a ten-thousand-year destabilization, that's a little worrisome. A lot worrisome. Timescales that are longer than we have experience with are going to matter.

Even at the surface, I'm curious how much we can see already. We hear a lot about overfishing, dead zones, and plastic in the ocean, and this conversation makes me wonder if we can see changes in ocean chemistry. In the context of this project and the reenactment of the dare, I've been reading about other reenactments. One of them was the Kon-Tiki Expedition, in which seven Norwegians crossed the Pacific on a raft in 1947 to prove that Polynesia could have been settled by travelers from South America. Thor Heyerdahl's account of the abundant sea life is just amazing—glowing plankton blooms, fish jumping out of the water and into the frying pan, huge sharks and whales swimming alongside the raft. Would the ocean still look that way if one was to make such an expedition today?

That's a really good question. Last time I went to sea was ten years ago, but I can certainly say, from my time on ships in tropical regions, that you pass through algae blooms with phospholuminesence every night. Or a pod of flying fish will come along and fly up onto the deck and make a mess—and then you realize, wow, there are thousands of flying fish all around us. Or you'll be sailing along and see a bunch of dolphins playing in the bow wake of the ship. When you see them, you know for *sure* that they're playing; there is no reason for them to be following the ship, diving in and out of the wake. Ten or fifteen dolphins will come along and surf on the wake—when one comes off, the next one comes on, and they circle back around to do it again! Spending time on the ocean, you would be amazed at what is out there. But as we see on the news all the time, you would also be amazed at the amount of plastic, garbage, and things that Heyerdahl never would have seen. But then, I see this as kind of surface—the skin of the ocean that we can see pretty easily.

We can't see that the chemistry of the ocean is changing. That's an interesting contrast. The ocean is so big, and people's regional experiences are so highly variable. But the chemistry is not regional. Scientists have measured it in Hawaii, and it turns out that the same pattern exists in Bermuda, Tahiti, and the Indian Ocean. The acidification that we've started to measure is everywhere; it surpasses whatever regional conditions people are encountering. The spatial aspect of the ocean is another thing we can't wrap our heads around very well.

So if I try to imagine that surface story layered onto the deep ocean, where does that half century fit into the thousand-year cycle you mentioned?

Well, here's another scale story. It's partly a statistical argument to say that a single molecule of water from the North Atlantic takes between eight hundred and fourteen hundred years to circulate through the entire ocean and back to the North Atlantic. The ocean mixing time is around a thousand on average, and it's kind of a conveyor belt. You can start anywhere, but if we start near Greenland, that molecule of water melts off a glacier and gets cold and salty enough to sink down to the bottom of the North Atlantic. Then it traces its way down past Iceland; it hugs the coast of North America, down south of Bermuda, comes around Brazil, all the way to the Antarctic. Then that parcel of water does a fabulous thing. It goes through the Antarctic Circumpolar Current at the bottom of the ocean, and after it goes around Antarctica, maybe it gets shot off into the deep ocean in the Indian or the Pacific and eventually it finds its way up to the North Pacific.

Through that process, the parcel of water also goes through what's called upwelling, which brings deep cold water back up to the surface. It's like a Möbius strip driven by gravity. Cold salty water is dense, and so it sinks and is then pushed along by water sinking behind it. For comparison, ocean currents are measured in Sverdrups (Sv); one is equal to 1,000,000 cubic meters per second. That's a huge amount of water. The Gulf Stream, a fast current that moves ships easily from North America to Europe, is 50 Sv, while the deep ocean current is 1 Sv—a slow, steady flow crawling along the bottom of the ocean.

And how much of it evaporates and goes up into the clouds?

All along the surface flow, there's evaporation happening, especially through warm tropical areas, and that's what drives the temperature of the surface of the earth. The Gulf Stream and the North Atlantic Current carry that warm tropical water, and that warm tropical water evaporates into atmosphere to help thermo-regulate the northern areas of the planet. We retain our nice habitable environment because of heat flow from the tropics to the poles. This is where the mass of the ocean versus the mass of the atmosphere comes into play. The mass of the atmosphere is tiny compared to the mass of the ocean, and so the number of molecules that get transported into the atmosphere and that are driving this whole surface heat flux is trivially tiny compared to the slow, relentless circulation of the ocean.

Wow, I've had my head in the clouds again…Climate change is something I imagine happening in the atmosphere. But this is just because we recognize it there first—because the atmosphere changes rapidly enough for us to notice it in our everyday life. Ignoring the register of climate change in the depths of the ocean seems like a wild oversight. But what about the geosphere? There's a theory that climate change triggers more volcanoes, earthquakes, and tsunamis by changing air pressure and water circulation patterns that put pressure on the tectonic plates. On land this makes sense, because some volcanoes literally have a glacier sitting on top. But what about in the ocean?

We do know that climate change is melting ice caps, and there is some evidence from Iceland that the reduction in pressure from melting ice can increase the frequency and size of volcanic eruptions. It's absolutely true that if you un-lid a volcano by releasing the pressure of ice—ice is heavy—you can change the pressure balance for that system. We don't know, though, just how carefully poised things like ocean plates are. But I'll tell you this—the fact that hurricane-scale storms are more frequent, and more intense, is absolutely tied to climate change. You have to have warm water and enough evaporation at the surface to get enough water into the atmosphere to start a hurricane. In the tropical Atlantic and tropical Pacific, where cyclones and hurricanes form, we know we have warmer surface water, and we know it's because CO2 in the atmosphere is raising the temperature of the planet. If hurricanes are influenced, then I can't say for sure that earthquakes and tsunamis aren't either.

This goes back to the problem of what we perceive, as scientists, as a lack of information. We don't know enough about what triggers volcanoes and earthquakes in the absence of anthropogenic climate change. Half the problem with studying climate change is that our modern observational technologies and understandings of these processes barely grasp what happens on an unaffected earth. We don't get to study the unaffected earth. We have to study the earth we have.

I guess you'd have to wait an awfully long time for enough data to accumulate before you could detect a trend. We're already racing to keep up with climate change in the atmosphere—to understand it faster than it is happening—and the notion that we have to wait for more data is often invoked as a delay tactic. If these kinds of feedback loops between the atmosphere, the hydrosphere, and the geosphere are plausible from a scientific perspective, and it sounds like they are, they're interesting from a literary perspective—at least, fertile territory for science fiction.

We're way out into the unknown, and for now the unknowable, so let's turn back toward safer shores. This project emerged from an interest in how exposure to unfamiliar environmental conditions impacts people, whether by provoking fear, curiosity, or inspiration. We've been talking about vast scales and invisible processes, and what I want to ask you now is, what happens when you go out into the field and encounter these processes directly? What changes when you go there? What does fieldwork mean for artists and scientists?

> It's one thing to know, in concept, that I'm working on a river. But it is not the same to go to the site, the same river, over and over again. The world is dynamic in ways that I need to see to understand. There may have been a landslide that inverted the river's flow; or because of a landslide one hundred years ago, the river might now bend around a large rock and that bend, which has a slower circulation, means algae might build up or a different chemistry might emerge in that spot...The physicality of the environment matters, even to a chemist. If someone brings me samples from an environment that I haven't experienced, I always feel like I don't *really* know what I've got. Does that make sense? They can give me all the information. They can tell me what the temperature and the weather was like, and what they saw, but there's a piece of me that needs to experience where the sample comes from firsthand.
>
> I don't think this is unique to me, but maybe it's because I'm outdoorsy and I've evolved this feeling of the earth as a system that I'm as much a part of as my samples—and if I'm not out there collecting samples then I feel like there's something missing. I don't want to go too far down this road, but for me there's something spiritual

about being outside, in the place that I'm trying to study and understand. The act of collecting samples in the environment matters to the way I do my science.

The way you describe fieldwork reminds me so much of *A Feeling for the Organism*, the biography of Barbara McClintock, and her description of the intuition she had for corn DNA. It might seem like that kind of empathy or intuition would just be for organisms, life forms, but my sense is that maybe it's a more pervasive form of orienting oneself, a kind of imagination and intuition for your subject matter that isn't limited to organisms. It could be for chemical processes or rates of change—

Here's a tiny example: The human sense of smell is extremely sensitive to sulfide—the smell of rotten eggs. For a long time, I studied marine sediments at the seafloor, which often produce sulfide. Collecting these samples, I can smell something about the sediment that tells me about what its chemistry is like. I'm not super well calibrated, but the visceral smell and feel of deep-sea sediment tells me something. Color also matters. Deep-sea sediment that has a lot of carbon in it is a brownish color, and once all that carbon has been burned away or oxidized by organisms, the sediment turns gray. I can't tell exactly how much carbon is in the sediment by looking at the color, but I can see there's biology happening in it—I just know there's life there.

I don't know how this would work for someone who's a theorist or a laboratory scientist, but I'm certain they'd still have the same kind of visceral impressions. This is what makes me excited to get up in the morning and think about the world. I think we, scientists, particularly academic scientists, can't do much else with our lives. We are compelled to do science the way an artist is probably compelled to paint. I'm not sure that people realize that that's part of what makes us tick. There's this perception in the world that scientists are dry and stodgy, that they're not funny, and that they just think about data and don't get excited about stuff—but it turns out we do. It's as much a human endeavor as anything else.

Recently I took a group of new students on a creek walk in the mountains in Arizona. You ask people what they see and, at first, they're afraid to tell you because they don't know what they see. You prod them a little bit. They say, "I see water; I see rocks," and then they start to recognize that they see evidence of live animals—they see that deer have been walking through the river, that there are plants growing in the water, or they flip over a rock and see bugs. It takes a while to be able to describe what they've seen. If you had asked the same question in the classroom, they would never have dreamed about the richness of the environment that they saw. People have a hard time recognizing and interacting with the complexity of a thing even when they're experiencing it. When you go to a site, you have to tell a story about what is actually there instead of what you imagine is going to be there. I do that with data and with my experience of the place where I got the data. I don't think it's that different from someone who paints *en plein air*.

It's often said that the reason there isn't more concerted action on climate change is that most people don't have any direct experience of it. We don't do anything until the roof falls on our heads. There are lots of reasons for this, but part of the story seems to be connected to these issues of being able to recognize what's going on right in front our eyes, on the one hand, and to imagine these long timescales on the other. Can you tell me a little more about how you think these issues need to be addressed and what you're doing with PlanetWorks?

I realize that it's a very First World luxury to be able to reflect on any of this. People are starving; they don't have water; they're fighting. The thousand-year ocean circulation doesn't matter to these people, and that's OK. If you're a subsistence farmer, climate change matters and is going to affect your crops. But in reality, whether or not you have enough rice for tomorrow matters more. This is why you have to look out for greater good. Most of the people who are going to suffer at the hands of climate change are in no position to do anything about it, and they have very little say in how the system is going to evolve. For those of us who do see and recognize the scale of the system that we're working in, there's a huge amount of responsibility that comes with that knowledge.

To bring this back around to PlanetWorks—understanding how the oceans operate and how they respond to a changing climate is one (important) part of understanding what the future of our planet might be like. But if we want to achieve a thriving future for all people (and I think it is a safe assumption that we do), we might reach a point in time where humans decide to intervene in the planet's climate system. How we might do that is not just a science question—it's a question of values, culture, governance, technology, and, of course, some science. What can we do? Who decides that we should do it? What are the potential impacts of those decisions? This work must involve humanists, scientists, artists, engineers, ethicists... all modes of endeavor. PlanetWorks is a collaborative, cross-disciplinary initiative that aims to learn how to make thoughtful decisions about the coupled human system–earth system. The idea of intervening in the earth system might be frightening to some, and perhaps it should be, but it will definitely be scary if we decide to intervene without first trying to understand the immense range of implications of that intervention.

If you have the luxury of noticing, then that luxury comes with the responsibility to actually pay close attention and do something with that awareness.

Artists and scientists are compelled to respond.

Last thing: temperatures at the poles were terrible this winter. It was maybe even a "year without a winter." I understand you've written a paper on a technological system for refreezing the Arctic. Although it doesn't sound as scary to me as mimicking the effects of a volcanic eruption, it does sound like the sort of "Frankenplanet" stuff that worries some people about geoengineering. Can say something about why we might want to consider these options?

> The last three years have been the warmest in recorded history. We (and here I mean humanity) may reach a stage where we decide that we must cool the planet and intervene in Earth's climate system. This isn't a far-reaching concept—we warmed the planet when we decided to burn fossil fuels. We could now decide to use our ingenuity to figure out how to cool the planet. I've been working with a team of scientists and engineers on a concept for rebuilding the sea ice that is lost due to Arctic warming. Ice in the Arctic reflects sunlight back to space. This is a key climate feedback that keeps the earth cooler. Our hope is that by thickening this ice in the winter, we might delay or prevent the loss of ice in the summer. Perhaps we could utilize wind-powered pumps to spread water over the ice (a bit like building an ice rink in your backyard) or maybe it will be some other yet to be discovered technology. Regardless, we need audacious ideas for climate intervention and there are not many options on the table right now.

# Notes from Tambora: or, Planetary Persuasions

Gillen D'Arcy Wood

*In May 2011, I traveled to the Indonesian island of Sumbawa to climb Mount Tambora, or what was left of it. I had learned about Tambora's 1815 eruption—the most powerful on Earth in ten thousand years—in an atmospheric science class I took while on sabbatical from my English department, and had resolved to write a book about its effects, which had gone largely unrecorded or ungathered. The raw physical nature of the subject convinced me the book could not be written solely from the safety of the archive. So, I traveled the globe—in the path of Tambora's plume—looking to piece together the fragments of this epic story. I brought one book with me to Tambora: Jane Austen's* Persuasion, *written in 1816, the cheerless Year Without a Summer that followed Tambora's explosion. And I brought a journal.* Frankenstein: or, The Modern Prometheus, *the most famous novel of 1816, was already key to my plans for my Tambora book, but I wondered, on that trip, what insights a volcanic pilgrimage half a world from Austen's native Hampshire would bring to her deathbed novel of 1816 and to our own much-altered post-Tambora world.*

On the bone-jarring road out of Bima, capital of Sumbawa, I first see Tambora as a long gray giant under a blanket of cloud. One pothole too many, and our SUV bursts a tire. I stroll across the pasture by the road in view of the sea. This is Tambora's new coast, Tambora's lava landfill. Its eruption two centuries ago blew the mountaintop into the sky. A volcanic haze drained the world of heat and color, unleashing three years of wild weather and biblical suffering. I try to imagine plumes twenty-five miles in the air, trees like burning javelins into the sea, but I fail at that.

It's too peaceful here to conjure up apocalypse. Distant cowbells drift across the pasture, like a faint gamelan orchestra. Seventy-two hours ago I was shoveling snow in a driveway in Illinois. The teleconnections of modern travel are like the teleconnections of climate. Fist-size rocks—black and granular—are strewn all about me. Volcanic debris. This is not a tourist site. Souvenirs of a forgotten disaster are worthless. So it is the easiest thing in the world for me to reach down and pick up a rock, as if it had just fallen there yesterday, the terrible night of April 10, 1815, when thousands perished in a fiery hour. I put the rock in my pocket.

In the summer of 1816, Jane Austen battled to finish her last novel while her body rebelled. She suffered constant fevers and broke out in spots. Her head, her jaw, throbbed. She carried a small cushion to press on her cheek, ineffectually, against the pain. There's no record of her treatment with leeches. She couldn't walk and could barely hold her pen. Austen's heroine of 1816, Anne Elliot, is an avatar of the author's fatal wintry illness. Twenty-seven years old when *Persuasion* commences, Anne has "lost her bloom early." According to norms of the marriage plot novel, she is effectively dead. *Persuasion,* like the other great Tambora novel, *Frankenstein: or, The Modern Prometheus,* centers on the elaborate reanimation of a corpse. Anne's second chance with Captain Wentworth is Austen's fantasy of the "second spring" she herself will be denied. The summer after the Year Without a Summer, Jane Austen is dead.

*Persuasion* and 1816. Summer denied was like love denied. *Frankenstein*'s monster enacts the stormy destruction of 1816, while Anne Elliot is the storm's victim: a hollow woman, a stuffed woman, headpiece filled with straws of regret. Anne almost never talks. Her listless body, like her creator's, is empty of hope, the pale vessel for a season that never was. The summer of 1816 never did happen, except calendrically. A summer in deep winter masquerade. No sun, no warmth, like the end of days. Those who lived those weeks of rain underwent a peculiar unraveling. The mind insisted on summer, like a wish list, a mental catalog of pleasurable expectations. But the body only suffered, in a nutritional vacuum. In July, Austen commiserates with a Hampshire farmer that "it is bad weather for the hay." He comforts her with the reflection that "it is much worse for the wheat." How much hunger subsists in that passing exchange? Empty bellies across

Europe and Asia. In 1816, the starving peasantry ate cattle feed. Hollow men, hollow women, stuffed with straw.

Tire changed, my driver looks up at the mountain for the first time. The name Tambora, he says, means "to appear and disappear." A word shard left over from a language the volcano itself destroyed, in 1815. There and not there. A volcanic name uncannily tied to our perception: the earth we inhabit that is both here and not here, at our convenience; a volcanic eruption that is both a magnitudinous event and a giant hole; an author full of words but empty of person. We know so little about Austen and her last illness. A hollow woman, in other words.

Her sister Cassandra made a small inferno of Austen's letters. The first committed to the flames were those that described—too graphically—the full symptoms of her 1816 illness. Austen's bodily fluids were not for posterity. If tubercular, then gobs of phlegm. If cancerous, then bouts of vomiting. Austen's body erupts in protest against the climate. As she wrote to her brother in July 1816: "It is really too bad, & has been too bad for a long time, much worse than anybody *can* bear." This is Austen's body talking through the medium of weather: "Oh! it rains again; it beats against the window."[1] The pain, the pain. For Cassandra, it was vital to preserve the image of a cheerful, resilient sister, who died uncomplainingly after a short illness. But Austen out-pointed her loving censor by figuring her distempered body through the storm skies of 1816: "I begin to think it will never be fine again." Sure enough, she never was.

The climb is kind at first, among the coffee plantations built by the Dutch in the days of the spice trade. Slaves labored here for two hundred years under their Chinese overseers. Then we enter the jungle. My guide, Ma-cho, carries a machete and stops every few yards to thrash at ferns and nettles and pick leeches off his skin. Ma-cho has climber's legs, with muscles like sailor's knots. Next to him, mine look like greasy chicken thighs on a lunch tray. How many Indonesians does it take to transport an American academic up a volcano? I can report the answer is five, conservatively. While I reapply insect repellent and take a

1 *Jane Austen's Letters*, ed. Deirdre Le Faye (Oxford, UK: Oxford University Press, 2011), 329–330.

puff of albuterol, my companions look at me pityingly, a helpless Westerner in the jungle, clutching his little medicine bag. I sip morosely on cups of very sweet tea.

Ma-cho takes conversation with clients seriously, as a professional service. He practices sustainability management, though he enjoys no official title. In his village on the nearby island of Lombok, he advises his neighbors on the disposal of their ever-rising tide of plastic waste—bottles, bags, tubs, wrapping. At Ma-cho's urging, his village no longer burns its trash in piles by the road. Instead, they have dug pits at a distance behind the store where they put all their empty plastic. When the hole is full to the brim, they set fire to it. Then they dig another hole. Another hollow in the ground for more hollow stuff.

In the bitter chill of the Year Without a Summer, late in July, Austen rewrote her ending to *Persuasion*, as if alert to the historical lie of happiness. Austen rewrites the conclusion, while the heroine Anne rewrites the novel itself. First kiss barely enjoyed, she insists, to her uncomprehending lover, that she was right to refuse him eight years ago, that their death-like suffering *had to happen*. Anne makes no sense, except in the insistence on suffering. In Anne's rogue reading of her own story, persuasion is coercion, suffering inevitable, and "choice" the offer you can't refuse. She insists on a deterministic world, when we have fallen headlong in love, as we must, with her free will. Historical romance craters beneath our feet, while Anne emerges as a true heroine of 1816. The lesson of the Year Without a Summer is not human self-determination—sweet reward of the liberal novel—but of people, *en masse*, as the collateral refuse of planetary persuasions.

On the jungle slopes of Tambora, it has been raining for five hours when two Swiss climbers appear at our sodden camp in the jungle, where monkeys are screaming in the trees. One of the Swiss men is compact and sinewy, the other bearded and bearish with a kerchief tied across his forehead—like a cartoon duo. Ma-cho is quietly appalled. The Swiss set out late, with no tents or food for their porters. Now darkness is falling, it is pouring with rain, and the village men face a wretched night sleeping with the leeches on wet grass in the rain. "My heart is broken for them," says Ma-cho over and over. He invites the porters into his tent and feeds the Swiss, who have not bothered to pack food even for themselves. It means less for everyone. They do not thank him. With uncanny Swiss precision, the two Europeans have managed to reenact the history of East Indies colonialism in a single afternoon. It's the old story of careless Europeans, and their aerobic hubris, treating the world as their playground and indigenous peoples like trash. They speak only to me. And how friendly and civilized they are. And how slavishly grateful I am for their polite interest in my book (the shame will come later). But neither are the homeless porters grateful to Ma-cho. Instead, they talk and joke inside his tent all night, keeping him awake, and when at last they fall asleep, they snore uproariously. Ma-cho does not reproach them, though he burns with resentment. Meanwhile, my whole body aches from the climb. The rain has soaked through every last piece of clothing I have. I spend the night killing leeches.

From the magmatic chamber of Tambora, an island's worth of gas and molten rock vaulted into the sky, leaving a hollow about half a mile deep, four miles wide. A hole you could fill with Tambora's dead. But the words for these dead are missing, like Austen's burnt letters. We build history from texts and tales, not from magma, sulfate aerosols, and deadly microbes, the inhuman stuff of disaster. So the crater is left empty, as if raw earth cannot be made story. Like Jane Austen coughing up phlegm, or vomiting blood, certain stories resist writing because they are too much of the body. Tens of thousands dead of starvation, of fluorine poisoning, of typhus. Millions from cholera. Think of Tambora as a continuum of gaseous fluids, passing from magma to sulfur to rain to human phlegm, shit, and blood, with only the thinnest membrane of separation. Think of Lear's "bare, forked animal" in the storm on the heath, then subtract the poetry. That was Tambora. That is climate change.

If seasonal affective disorder were a novel, it would be *Persuasion*. Only in 1816 could bad weather be made into art. Except perhaps in 2016 and beyond. In Austen's letters of 1816, she insists to family and friends she is getting better, even as she is dying. Summer and winter in one. In *Persuasion*—Austen's novel of dissimulation—we have two novels, with two endings and two heroines. Anne's pale, wasting body—internal figure of her creator's decline—stands in for the hollow of history. Her internal monologue insists on the hopeless status quo while her body registers historical change, her connection to the biophysical world. With the return of desire, her color returns. She grows pretty, not like her former self but like a different person. Mr. Elliott lusts after her in a chance encounter, as if she were not his plain spinster cousin. Her father asks what facial cream she's using, as if she were a Bath socialite. She's an alternate heroine in an alternate summer.

Tambora is not Ma-cho's volcano. His is Mount Rinjani, across the bay on Lombok, which he climbs thirty or forty times a year as a professional mountain guide. It was Rinjani that exploded in 1809, first cooling the globe and setting the stage for the climate chaos of Tambora. Rinjani and Tambora together ensured the 1810s as the coldest, stormiest decade of the last millennium. Near the summit of Tambora, Rinjani is visible across the sparkling blue sea. During its last eruption, in 1994, when Ma-cho was a boy, he fled down from the mountains to the sea with the rest of his family, stupefied with fear. They lived on the beach for a week. Ma-cho is pleased for me that it has stopped raining and that I will see what I came inexplicably to see, but when the clouds lift, and he glimpses the fearsome outline of Rinjani in the distance, he instantly forgets Tambora and lets out a whoop of joy.

The rain beats on the window at Chawton as Austen fights the pain to eke out the last scenes of *Persuasion*. How to reunite the lovers? Of course, it must be raining, and the comedy must turn on an umbrella. At this point, at the lovers' maximum estrangement, Anne's repudiation of desire is a mind-forg'd manacle not easily loosened. Her sexual despair rings like climate denialism, the great corporate sophistry of our age. Both despair and denial weaponize modes of reason in all-out assault on the senses. In the shop on Milsom Street in Bath, Anne insists to herself that she will walk the few steps to the window to see if it rains, *not* in hopes of seeing the captain. To reach the window, against such crosswinds of cognition, is an Olympian feat. She must traverse the shop floor while denying the floor, her body to cross it with, that the window exists, the street, the rain, and even that the captain is physically in Bath for her to see him. Why then do we shake our heads at climate denial, when Austen demonstrates a cognitive equipment so supremely adapted to delusion? The lesson of 1816 is the rogue lesson of *Persuasion*. Without sun and desire, life withers. Climate is written on the body, even as the mind rationalizes its freedom from earthly constraint.

In the night, the earth reverberates beneath our tents. A deep sound almost beyond hearing, like from the well of a dream. Two weeks later the government will order evacuation of the entire peninsula. We wake to a light, sulfurous mist. Ma-cho says, "I hope today is not the day." Kept awake by his ungrateful Sumbawan tentmates, Ma-cho has spent the night worrying about how to tell me that I've come all this way for nothing, that it is too slippery and dangerous for us to climb the last mile to the crater in the rain. But with the dawn, the clouds have cleared, and we make the climb into sunshine, careful to keep to the grassy side of the ridge, and away from the loose rocks down the ravine. At last, at the rim itself, the grueling, step-at-a-time torture of the ascent melts into make-believe. The caldera opens up like a great blue-green pearl from the mountain depths. Tambora is a mountain without a summit, a volcano with no cone, a giant cyclops head nursing its gaping wound. At the crater's edge is loose black sand; it slopes invitingly into the abyss. The wind is blowing hard at our back, toward the caldera depths. It plucks at my jacket, nudging me forward. Ma-cho draws a line in the dirt, ten feet from the edge, as if Tambora needs babyproofing.

Metaphors of melancholy were made literal in 1816. Gloom and despondency triumphed in a sunless world. For Austen herself, she will not live to enjoy the sun's return in 1819, that overdue "season of mist and mellow fruitfulness." She will never escape the gloom of 1816. So she plots an escape for her heroine, a fantasized return to fine weather and rude health. Like *Persuasion*, 1816 was a world of double stories, sliding doors, alternate endings. In conventional histories, the European armies of 1816 de-mobilized, flooding cities and villages with unwanted labor. Struggling to adapt to postwar conditions, manufacturing stalled and crashed. Wartime blockades lifted; nations jostled for trade advantage. Political and economic chaos. Hunger and suffering. To admit the intervention of an Asian volcano into these complex post-Napoleonic affairs would be to question human sovereignty in general and the self-determination of white, North Atlantic peoples in particular. So Tambora—the monster eruption—is refused admittance to scholarly history, just as Frankenstein's creature is barred to all civilized society.

At Tambora's rim, all is exposed, hollowed out. The aridity of a moon. That's the volcano experience. The earth has shed all vegetation, all pretensions of caring. It reveals its brutality, and we are like shells washed up. On Tambora's brim, death seems like a lovely thing. My fleeting distress is the meagerest residue of the historical trauma wrought by the Great Eruption itself, in which children went unburied, Mary Shelley saw monsters, and Jane Austen—hollowed out with pain—bore witness to her own dying through the pale gaze of Anne Elliot, her own creation, like the Creature haunting Frankenstein.

On the journey back to Bima, we stop at a tiny roadside station, crammed between the jungle and a swampy inlet. A bullock is stuck in the mud by the road. A girl serves us instant noodles in plastic tubs, silently under the watchful eye of her father. She is clean and pristine, while we are filthy travelers. Piles of plastic bottles and trash line the road; some of the piles are burning, sending plumes of smoke into the skies of Sumbawa. I know this obscenity of waste scandalizes Ma-cho, who has dug holes in his home village to put their fresh garbage. But he makes no comment to the girl. She, after all, must live in it.

The next day, I leave Sumbawa. The soaring cities of Asia and America pass by in a haze, like debris plumes of the built world. On the way, I see enough plastic bottles to fill ten Tamboras.

## Why Go?

Gillen D'Arcy Wood in conversation with Dehlia Hannah

DEHLIA HANNAH I learned about the eruption of Mount Tambora while reading about the governance of geoengineering technologies.[1] The article examined the prospect of deploying sulfate aerosol injection as a strategy for mitigating climate change; it began with a suggestive anecdote about the Year Without a Summer and the writing of *Frankenstein: or, The Modern Prometheus*. The volcano itself seemed a distant place holder, a causal mechanism setting into motion a more proximal literary and scientific history. Your book, *Tambora: The Eruption that Changed the World,* initiated a profound change in my imaginative geography, ultimately inspiring me to retrace your steps. When you began your research on Tambora, did you envision visiting the mountain, and how did you imagine this as part of your research?

GILLEN D'ARCY WOOD The pilgrimage framework emerged early: an ascent of the volcano itself, then the pursuit of its deadly plume across continents, from China and India north to Europe and the Arctic. A literal journey to parallel the journey of the historical imagination. Tambora itself, of course, was utterly special, as the tropical ground zero of the story. Mountains, by their nature, are iconic and lend themselves to mythology. But because of the 1815 eruption, Tambora possesses an aura equivalent to Santorini or Sinai. The enormous caldera—where the mountain peak *was* but

1 Jack Stilgoe, Matthew Watson, and Kirsty Kuo, "Public Engagement with Biotechnologies Offers Lessons for the Governance of Geoengineering Research and Beyond," *PLoS Biol*, vol. 11, no. 11 (November 12, 2013).

is no more—looms out of the mist like a time portal, where geological time and human history intersect. The experience evoked the sublime but was, in its way, anti-Kantian.

I felt no resolution of self in the aftermath, only the urgency of what the caldera came to symbolize for me: an unwritten history of human beings and the planetary motions that break the frame of our perpetual self-regard, a reality our pre-industrial ancestors were more alert to than we.

How does fieldwork or site visitation inform your notion of environmental and literary history?

From the outset, I wanted an embodied dimension to the Tambora narrative. The book bears witness to the suffering of millions, and to the pathos of their unremembered fate. Much of the book, too, concerns physical environments, and the raw effects of climate change across wide latitudes. This demanded, as I saw it, a greater level of authorial involvement than the scholarly modes I was used to. But I wouldn't want to endow my research methods with a retrospective false certainty. Much of it was a groping forward, led by impulse. Travelogue, memoir, ethnography: I toyed with genres and narrative modes but settled on a form of Romantic storytelling, inspired by the Shelleys, in which poetry, science, and history are all one and in which the intimacy of selves is always undergoing some revolution of scale, being translated into a thing astral or transhistorical, relevant to people and the earth in their broadest senses.

It was extraordinary to meet you in the summer of 2017 and learn that we had both visited Antarctica earlier in the year. What led you to Antarctica after this major work on Tambora? Did the visit to Tambora resonate with your recent trip to Antarctica?

The narrative scale demanded by Tambora—global, geological—became intellectually addictive. My other books have been mostly modern-metropolitan, even bourgeois-domestic, in their concerns. After my early career as a Romanticist, I had so utterly absorbed its tropes that I was no longer working as a Romanticist but a veritable Romantic, as such, drawn ineluctably to remote volcanoes and glaciers and all that physical furniture of the sublime. So in terms of scale, Tambora to Antarctica assumed a linear progression in my mind. But of course, the temporalities are radically different. *Tambora* worked well as a conventional disaster narrative—a specific violent event—with a long-lens climate change theme smuggled aboard, as it were. By contrast, for all its disintegrating ice shelves, Antarctica is a land without events as we understand them. We barely have a language to describe historical change as it unfolds there. The challenge, with both Tambora and Antarctica, is to render these places narratable, to bring them within a discursive horizon. Both are climate change projects designed to burst the presentist bubble through which we view the existential threat of global warming: the first as a phenomenon in historical time (1815–1818), the second in deep time.

What's the significance of retracing the steps of past real and fictional figures through environments such as these?

History is a collective mythmaking the style of which periodically undergoes revolution. In our revolutionary moment, the old anthropocentrisms appear suddenly inadequate to the stories we need to tell about ourselves in the century of climate change. Certain characters of the past—mere extras of conventional history—loom large as fellow travelers. I'm thinking of, say, Luke Howard, the pioneering meteorologist, or Li Yuyang, the Chinese poet—key figures in my Tambora narrative. Not only their existence but their *relationship* is made visible only through an ecological lens, at the expense of anthropocentric logic and its tyrannies.

How do you feel about activating Tambora as a historical and cultural site, for scholars, the local community, the tourism industry, and contemporary artists...?

> The "remoteness" of Tambora is not merely a perceptual artifact of North Atlantic privilege. Sumbawa, Tambora's island, is remote even to the Balinese, barely a hundred miles away. In other words, Sumbawa's remoteness is a function of poverty and underdevelopment. The rural villagers live mostly as they did decades ago, even a century ago. Sumbawans of the older generation still ascribe the island's misfortune to the Tambora eruption. That the "discovery" or commercialization of Tambora might offer a partial economic salvation, or recompense, is an irony still difficult to envisage, despite the attention of the bicentenary (which inspired the Indonesian government to rebuild the road from the capital Bima). The Tambora bicentenary conference in Bern, Switzerland, said it all: only two of the dozens of conference speakers had actually been to Tambora. It's a humbling reminder that our research agendas—in the sciences *and* humanities—so often reflect the parameters of the known, so rarely venture "off road," where there's no infrastructure, no established funding stream, no invested audience. Will Tambora attract these? Will we look back at the bicentenary, and the *Tambora* book, as the beginning of a Tambora industry, scholarly and commercial? It's too early to say. For the time being, a Tambora pilgrimage still requires a highly motivated pilgrim.

Ultimately what I want to ask is: Why go? How would you describe the significance of visiting a distant/extreme/inaccessible place that you know so well from textual and other sources?

> Doesn't the twenty-first century feel different from the late twentieth? It does to me. In this time of unprecedented physical change in the world—its weather systems, its biota, its landscapes—texts appear increasingly unreliable, or partial. Some more immediate act of witnessing is

required. Pilgrimage is one alternative mode, and scientific literacy another. I'm a humanist who believes that the critique of data and the scientific method misses the point. Graphs and figures are only a snapshot of the world—like an image or metaphor. Today's tsunami of research in the physical sciences is raw material as rich as Balzac's Paris, waiting to be narrativized, to be well wrought as a new mythology of the earth. Without fresh data, our ecology declines into elegy, the jeremiad, and the circularity of critique. Ask yourself this question: Why does so much environmentalist writing sound the same when the earth is changing so much, so quickly?

Autumn passed thus. I saw, with surprise and grief, the leaves decay and fall, and nature again assume the barren and bleak appearance it had worn when I first beheld the woods and the lovely moon. Yet I did not heed the bleakness of the weather; I was better fitted by my conformation for the endurance of cold than heat. But my chief delights were the sight of the flowers, the birds, and all the gay apparel of summer; when those deserted me, I turned with more attention towards the cottagers. Their happiness was not decreased by the absence of summer. They loved and sympathized with one another; and their joys, depending on each other, were not interrupted by the casualties that took place around them. The more I saw of them, the greater became my desire to claim their protection and kindness; my heart yearned to be known and loved by these amiable creatures; to see their sweet looks directed towards me with affection was the utmost limit of my ambition. I dared not think that they would turn them from me with disdain and horror. The poor that stopped at their door were never driven away. I asked, it is true, for greater treasures than a little food or rest: I required kindness and sympathy; but I did not believe myself utterly unworthy of it.

—Mary Shelley
*Frankenstein: or, The Modern Prometheus*

# Listening for Seasons: Reflections from a Humanitarian Worker

Pablo Suarez

## I.

"The winter advanced, and an entire revolution of the seasons had taken place since I awoke into life," recalls the creature made by Victor Frankenstein, as he tells the tragic tale of his encounter with the family near whose cottage he found shelter.[1] In choosing the word "revolution," Mary Shelley most likely intended to convey the idea of a cycle, of recurrence. Yet we can sense other overtones in that word: a transformative insurrection, an overturning or radical change. An entire revolution of the seasons had indeed occurred two centuries ago when Shelley, stuck in a summer cottage near Geneva during the cold, rainy, summerless summer of 1816, first conceived the story of *Frankenstein: or, The Modern Prometheus*. Unbeknownst to Shelley, the previous year, halfway around the world in Indonesia, the eruption of Mount Tambora had sent gas, ash, and dust into the stratosphere, blocking sunlight and slowly cooling the planet. Over the following months, clothes froze on summer washing lines in New England, grain was in short supply in Britain, tens of thousands starved in China, and more perished as typhus spread across Europe. Frankenstein and his creature were born, in part, thanks to long evenings seeking warmth at the fireplace while Shelley and others shared haunting stories and conversed about the anomalous events unfolding around them.

A different revolution of the seasons is now happening around the world. And it hurts. We are used to the seasons anchoring frames of reference, offering reliable rhythms to the

1 Mary Shelley, *Frankenstein: or, The Modern Prometheus* (Boston: Cornhill Publishing Co., 1922), 173.

lives and livelihoods of subsistence farmers, fishing villagers, and so many others in communities relying on the annual dance with nature. Social life is choreographed around securing sustenance in harmony with seasonal ebbs and flows, around anticipating the risk of extreme events so as to prepare for the worst. As a researcher on climate and disaster turned humanitarian worker, I've heard countless African rhymes about the right time to plant and the best time to sow and celebrate; I've enjoyed Caribbean poems that fishermen recite to explain when to get ready for hurricanes and when to expect less threat the rest of the year. North of Ouagadougou, Burkina Faso, pastoralists speak of a season named after sandstorms. In the Pacific, navigators read the trade winds with the same precision that others feel listening to a metronome. Yet during the last decade, in my travels to more than sixty countries with the Red Cross Red Crescent Climate Center and other organizations, one story consistently emerged from very diverse voices: there's a new wickedness in the pattern of rains, winds, temperatures, and all related tempos experienced locally.

Strange things happen. Indeed, to paraphrase Aristotle, *it is in the very nature of probability that improbable things will happen.* Now, in a changing climate, Murphy's Law has expanded its jurisdiction. And when something does go wrong, it is my colleagues who have to rush to give a hand. Disaster management teams simply can't keep up. We need to rethink the climate risks we face. We have to reimagine who we are. At a time when artificial intelligence and other technologies are expanding our ability to see, think, and act, it seems that we're immersed in an ongoing, colossal failure to notice some of the environmental changes going on around us.

This new era is defined by us, humans, as a force of nature. We've been reshaping the land. We've become the unintentional alchemists of the oceans and the air—reengineering the very rhythm of our seasons. People and their calendars used to be in sync, harmoniously orchestrated and organized—so predictable as to be ingrained culturally and linguistically. Now our changing climate is rendering that traditional knowledge useless. Elders are losing respect and losing face because their forecasts are no longer trustworthy. We have robbed the world's villagers of their familiarity with the seasons by dissolving the groove of energy and matter at a planetary scale. We have gravely deprived them of the dignity of understanding and owning their feeling of home.

Two years ago, the Red Cross Red Crescent issued an invitation to museums and institutions where art is seen as a public benefit. It asked: "How can [we] work together for inspiration, reflection, and debate on climate issues? How can we help accelerate action by influencing culture? Can we mobilize the power of humanity to address climate risks, through innovative uses of art?" *A Year Without a Winter* offers a way of bringing these observations together in a compelling global story of many local changes. The project will surely change the way I draw on art, design, and games to help people understand and address the humanitarian consequences of climate change.

## II.

My grandmother, a risk management genius, lived to 106 years of age. Her younger brother died during the 1918 Spanish Flu pandemic that killed tens of millions. I grew up in her home in La Plata, Argentina. There was always a time when arriving home, whether from kindergarten or college, meant returning to the fragrance of freesias, which she had lovingly planted in the sun-facing terrace, under the kitchen window. I am still able to relive that magic: their aroma resides in my memory—I can evoke the details of that scent even though I left my hometown over two decades ago. But the scent was no longer there the last years I visited Nonna: when I asked why, she said that the terrace was getting too hot—her flowers were baking before they could blossom.

Of course, not only is the scent of freesias under threat given the reconfiguration of seasons. With summer conditions lasting into late October, the *New York Times* is now talking of "hotumn," with reports of carved pumpkins melting before Halloween and questions about whether Antonio Vivaldi needs to write a new concerto.[2] A new language is needed. In northern Ethiopia,

---

2 Reggie Ugwu, "'Hotumn' Takes Hold as October Temperatures Soar," *New York Times*, October 24, 2017, https://mobile.nytimes.com/2017/10/24/nyregion/warm-autumn-weather.html.

subsistence farmers speak of needing food aid due to a natural phenomenon they can't name—there isn't a word in their language for hailstones so big that upon impact they destroy crops. In Mozambique, the seasons are changing, and farmers don't know when to plant; the consequence is hunger. From around the world, my colleagues at the Red Cross Red Crescent share stories like these. We need the courage to learn new ways of narrating these conditions, because the unprecedented is becoming the new (ab)normal.

It is clear that in order to fully link science, policy, and humanitarian practice during this emerging epoch of humans as force of nature, we need to radically change how we envision plausible futures. We need art, design, games, and other innovative approaches for smarter learning and dialogue. As the lie that tells the truth, art can illuminate what we refuse to see right in front of us. Art can change the boundaries of what's perceivable and what's doable, connecting the infinitesimal to the infinite and motivating us to collaborate, chart new courses, and start new things.[3] As the purposeful specification of relationships, the field of design offers numerous lenses to blend and balance the impulses of control, autonomy, and cooperation—all things we confront when interacting with one another and with nature.[4] Lastly, as playable system dynamic models, games can help us experience the complexity of future risks.[5] Through gameplay, people combine the concentration of analytical rigor with the intuitive freedom of imaginative and artistic acts, capturing the essence of real-world systems and allowing for safe and rich explorations of how those systems could be changed. In the words of SimCity creator Will Wright: "Games amplify our imagination, like cars amplify our legs, or houses amplify our skins."

---

**3** See my piece "Climate Risks, Art, and Red Cross Action: Towards a Humanitarian Role for Museums," *L'Internationale*, November, 17, 2015, http://www.internationaleonline.org/research/politics_of_life_and_death/47_climate_risks_art_and_red_cross_action_towards_a_humanitarian_role_for_museums.

**4** Robert W. Keidel, *Seeing Organizational Patterns: A New Theory and Language of Organizational Design* (San Francisco: Berrett-Koehler, 1995).

**5** Janot Mendler de Suarez, Pablo Suarez, Carina Bachofen et al., *Games for a New Climate: Experiencing the Complexity of Future Risks* (Boston: Boston University, The Frederick S. Pardee Center for the Study of the Longer-Range Future, 2012), http://www.bu.edu/pardee/files/2012/11/Games-for-a-New-Climate-TF-Nov2012.pdf.

It is now possible to envision a "year without a winter." There are kids growing up today who will live longer than my grandma and, for them and their children, it is quite likely that the seasons will no longer consist of recognizable cycles but instead of puzzling phenomena, nuanced and shocking. Winters will not be what they used to be—in fact, they will probably no longer deserve to be called "winter" by the standards of current local experience. What new terms will we create? What should we expect after each solstice? Can we imagine what awaits us?

## III.

As disaster managers, what matters is our ability to prepare and respond rapidly to crisis situations, whatever their cause. Climate change and the resultant destabilization of seasonal patterns presents a clear hazard. But there are other, related, concerns. I met Dehlia Hannah at a geoengineering conference in Berlin, where scientists, engineers, policy analysts, artists, and activists converged to discuss the politics and technical feasibility of deliberately manipulating global climate. The figures of Prometheus and Frankenstein loom large in this discussion. As does the figure of Tambora, whose climate-cooling effects some seek to replicate today. Along with the impacts of climate change, the humanitarian sector should now be preparing to respond to threats to human well-being posed by future volcanic eruptions as well as geoengineering experiments.

The historical record shows major volcanic eruptions every few decades: the last one was in 1991. The impacts of the next climate-changing eruption may be felt at the global scale, and the humanitarian sector is, at present, completely unprepared for it (even for what questions to ask). Science can offer early warning signs between an observed explosive eruption and the likely cold, drought, and other extreme conditions expected several months after the eruption. Preparedness measures can reduce the losses anticipated by these models, such as those due to a reduction in the African and Asian summer monsoon. Yet the default scenario is that early actions will *not* be taken at the scale needed, especially in developing countries—unless targeted thinking and decisions emerge from interactions between scientists, humanitarians,

**Red Cross Red Crescent Climate Center, volcano preparedness flow chart for the project From Eruption to Humanitarian Action, grant proposal, 2017.**

donors, journalists, impacted communities, and many more, *before* the next big eruption. One must prepare to be ready.

We have developed a working group for the project From Eruption to Humanitarian Action aimed at understanding and preparing for the vastly under addressed risk of climate-changing volcanic eruptions—with, for example, forecast-based financing for early humanitarian action, an innovative approach that will be needed at the global scale if we want to act in time.[6] We know that, in order to mobilize the power of humanity, we will need not just money and people but also compelling narratives, inspired engagements, and creative interventions that can unlock new ways of thinking about our relationship to the environment and our agency within it. *A Year Without a Winter* is a step forward in this endeavor.

At the same time that we are preparing to respond to volcanic hazards and climate change crises, politicians, scientists, and engineers are seriously considering mimicking the climate

---

6 Erin Coughlan de Perez, Bart van den Hurk, Maarten K. van Aalst, Brenden Jongman, Thorsten Klose, and Pablo Suarez, "Forecast-Based Financing: An Approach for Catalyzing Humanitarian Action Based on Extreme Weather and Climate Forecasts," *Natural Hazards and Earth System Sciences*, vol. 15, issue 4 (April 2015): 895–904, https://www.nat-hazards-earth-syst-sci.net/15/895/2015/nhess-15-895-2015.pdf.

On lookout for climate impacts after #Calbuco eruption + bit.ly/1DDsleQ (Photo: Aeveraal via Wikipedia, 22/4)

12:31 AM - 23 Apr 2015

**The Red Cross Red Crescent Climate Center now monitors explosive eruptions that may alter global patterns of temperature and rainfall. The aim: to prepare for humanitarian action if extreme conditions are expected.**

cooling effects of Mount Tambora and other explosive eruptions through "solar radiation management" (SRM). Included in this climate engineering scheme is the proposal to inject a veil of sulfur dioxide in the upper atmosphere, producing essentially a light curtain to curtail global warming. Tests are under way. The proponents of geoengineering research (and its potential responsible deployment) I have met through the years, like their critics, seem genuinely motivated by good intentions. In our lifetimes, in the foreseeable future, winterless years, an iceless Arctic, and a fury of new seasons may call for desperate measures, forcing tough choices.

Whether unintentionally through greenhouse gas emissions or deliberately through geoengineering, altering the Earth's climate is an experiment in which every person on our planet is a test subject. We should do all we can to avoid having to face such a geoengineering decision—no one likes to be a rat in someone else's laboratory. But if humanity does have to confront this reality, we may, on that day, be asked to "trust the sulfur dust," and "act fast." That's Frankensteiningly concerning: who will care for the creature that is unleashed? There isn't a framework for dealing with the negative consequences that will likely affect those who have done little to cause the climate change problem and who will have played no role in concocting the "solution." Some of the language in geoengineering publications and discussions reveals potentially delusional assumptions about the rationality of the motivations, incentives, and triggers likely to shape SRM decisions and actions. Those who will suffer the worst outcomes need to be involved, especially given the risk of "predatory geoengineering," where recklessly self-concerned actions result in harmful consequences to others.[7] Here, the lessons of Frankenstein and the catastrophic effects of Tambora hold important lessons.

There are many unknowns in our path. In the coming decades, we are extremely likely to confront some exceptional developments—whether that means facing the aftermath of a catastrophic eruption, the last year with Arctic sea ice, the opportunity to switch to a carbonless economy in an electoral cycle, or the imminence of geoengineering deployment (either unilateral and predatory or concerted and meant as a force for good). We could face some unknown unknowns in the vast realm of plausible opportunities and threats. Whatever the case, we will need scientists, humanists, civil servants, farmers, and many others—greedy or selfless, powerful or vulnerable—to contribute to these big questions: What are the new narratives and images that enable appropriate and sound research, governance, decisions, and actions for living in the Anthropocene?

## IV.

When confronting a seemingly impossible and difficult task, my grandma used to say *"Difficile come pestar la Luna!"* ("As hard as stepping on the moon!"). She grew up at a time when our natural satellite was unimaginably beyond reach. Yet before I was born, there was a human footprint on the moon, and many more satellites seeing what we do—or fail to do. When analytical

---

7 Pablo Suarez and Maarten K. van Aalst, "Geoengineering: A Humanitarian Concern," *Earth's Future*, vol. 5, issue 2 (February 2017): 183–195, https://agupubs.onlinelibrary.wiley.com/doi/abs/10.1002/2016EF000464.

competence and creative talent join forces, we are capable of accomplishing amazing things. Well, now we must accomplish amazing things. I am grateful to *A Year Without a Winter* for the inspiration to think big, try hard, and do more.

It has been thoroughly exciting to see how *A Year Without a Winter* offers a way of thinking through these entangled issues. From casual conversations to executive meetings in the humanitarian sector, it is amazing to witness the collective response to the invitation to wonder what a winterless year would feel like, what it would cause, and what it would reveal about us. From interactive data visualizations in virtual reality to financial instruments for disaster preparedness planning at a global scale, these new ways of seeing allow us to play with the unfamiliar, enabling us to reimagine the space of possibility latent in our shared futures.[8] At the 2017 United Nations conference, COP23 Development and Climate Days, we designed and facilitated a session on *A Year Without a Winter*. We asked three hundred participants from around the world to imagine and share what it would be like to experience a year without their seasons—it was a deeply engaging exercise and it became a shared space for further conversation, helping to link knowledge with motivation for action. We have also facilitated this participatory approach to rethinking the seasons in various events, including the International Federation of Red Cross and Red Crescent Societies (IFRC) Statutory Meetings in Antalya, Turkey. The team in charge of global consultations to define the IFRC Strategy 2030 has decided to embrace "a year without our seasons" as a core dimension of our next steps. This is just the beginning.

---

**8** Pablo Suarez, "Virtual Reality for a New Climate: Red Cross Innovations in Risk Management," *Australian Journal of Emergency Management*, vol. 32, issue 2 (2017): https://ajem.infoservices.com.au/items/AJEM-32-02-05.

After failing art in high school, I couldn't have expected a professional career evolving in these directions. Sincere thanks to all the talented artists, designers, bridge-makers, and supporters that helped nurture this process, including Barbara Bulc, Cesarina and Maura Bertoncini, Carlos Mancinelli, Colleen Macklin, Francesca von Habsburg, Janot Mendler de Suarez, Jemilah Mahmood, Kathy Jetñil-Kijiner, Marco Castro Cosio, Olafur Eliasson, Pere Pérez, Pierre Thiam, Regie Gibson, Santiago Espeche, and Tomás Saraceno, as well as the inspiring team of partners and friends at the Red Cross Red Crescent Climate Center. The ideas presented in this paper are those of the author alone and do not necessarily reflect the views of the Red Cross Red Crescent, the IFRC, or its National Societies.

# An Invitation to Disappear: Postcards from Tambora

Julian Charrière and Dehlia Hannah

Dear Dehlia,

I'm writing you from a dam in the Swiss alps where I have been invited to create a project to be implemented on its crest next year. I just came back from a long hike around its lake—you would have loved it! The lake is surrounded by alpine peaks descending almost straight into soft, gray, silty water. On one side, the remains of what used to be a huge hanging glacier overlook the lake. On the other, there is a majestic waterfall, which fills the entire valley with the enormous sound of the infinite flux of water—an uncanny mélange of natural wonders and strange, synthetic elements. It's a vision of an artificial sublime. Everything here is a simulacrum, a construction: the lake itself, the waterfalls, even the feeling of the moisture on my face as I walk through the perfectly Swiss-engineered tunnel piercing the side of the mountain—a fully anthropogenic sight!

I just learned that the dying Giétro glacier overlooking this majestic landscape once blocked the valley at exactly the same place that the dam is positioned today. The glacial lake that formed behind it once frightened half of Switzerland, a result of the cooling effects of a volcanic winter in 1818. Is that not actually related to the very volcano in Indonesia you keep telling me about? And if so, is that not the strangest coincidence, a historical butterfly effect of some sort?!

Dehlia, I remember you saying we should go there to climb to the summit and hike into the caldera. I have a feeling this is a sign: The expedition is calling us—let's go!

Big hugs from Mauvoisin, Julian

1 Dessin non signé, attribué à Steinlen *Course à l'éboulement du glacier de Gétroz et au lac de Mauvoisin, au fond de la Vallée de Bagnes, 16 mai 1818.* Médiathèque Valais–Sion.

*A Year Without a Winter: An Invitation to Disappear* Dehlia Hannah, Julian Charrière, 2018

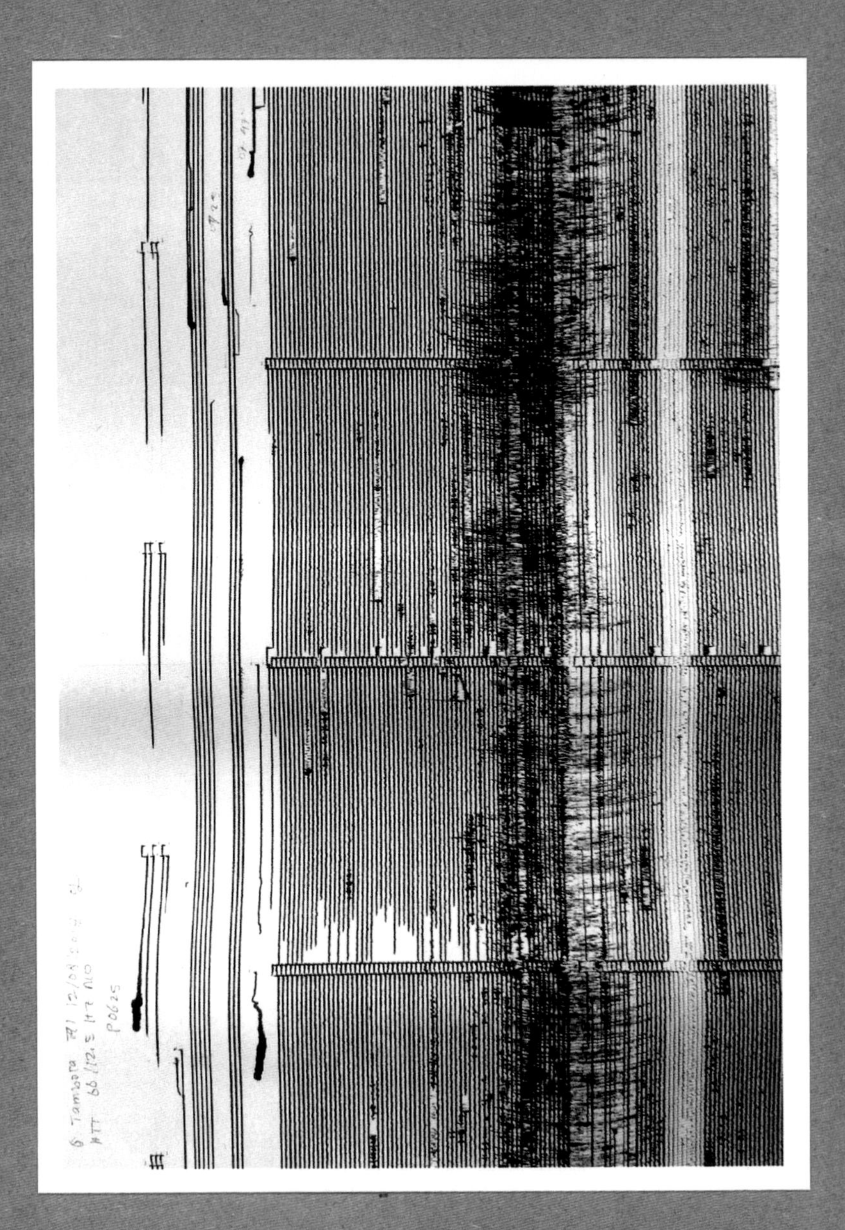

Dears,

We've arrived at the volcanic monitoring station in Doro Peti, a town of 600 on the northwest slope of Tambora, where we'll stay the night. Mr. Abdul Haris, the station head, will accompany us tomorrow—his first ascent to the caldera. Given that Tambora's last eruption was the largest in recorded history, we expected a somewhat larger operation, but it's a simple set-up: seismic rumblings are continuously recorded on an old portable seismograph and communicated to Indonesia's busy Directorate of Volcanology and Geological Hazard Mitigation. A framed poster shows a map of the magma chambers, evacuation routes, and a brief history of eruptions. It's been quiet since 1815, but in 2011 earthquakes and tall plumes of ash put the area on high alert. There's an uncanny mismatch between Tambora's outsized reputation and the treatment accorded here...

After an hour's flight and a six-hour drive to get here, we couldn't resist getting back on the road when our host offered to show us a baby volcano forming at the edge of the island. A burning, backfiring motorbike-ride later, we reached a crescent-shaped black beach, empty but for a small industrial facility. We plunged into the glittering blue water with our snorkel masks...the coral is the healthiest either of us has seen anywhere lately. Fat, fluorescent yellow and purple starfish glow stunningly against dark gray sand!

We returned to look through pictures and an old film given to us by Rik, who organized our trek, though he couldn't join us. There are incredible images from archaeological excavations started by Haraldur Sigurdsson, who came looking for the "Pompeii of the East" in 2004. There are bones, bits of modern pottery, and thick, carbonized tree trunks sticking out of cliffs along the beach, clearly swept down the mountain on fire. What really blew our minds, though, was the film: a 1939 Swedish documentary about a coffee plantation at Tambora's foothills during the height of colonial enthusiasm for "progress." Mesmerized, we sat in the dark (to avoid attracting bugs) and watched gigantic trees being felled by teams of thirty or more triumphant laborers—the last remaining jungle.

Our ability to converse with our guides is quite limited, so we're trying our best to read the landscape; it leaves us with so many questions...

Love, DH & JC

2 Doro Peti

*A Year Without a Winter: An Invitation to Disappear* Dehlia Hannah, Julian Charrière, 2018

Dears,

One speaks of mental clarity as the lifting of a fog, but our experience has been the opposite. We set off for Tambora this morning with Mr. Haris and three guides, piling into a jeep that must have been from the 1950s. The first leg of the trip took us part way up the mountain, a wild off-road ride over a landscape littered with huge black stones that look like they fell from the sky yesterday. Grasses and low trees have grown up around them. We wondered whether these slopes had recovered naturally after being completely stripped by the eruption, or whether they'd been burned and cleared more recently for crops.

After many hours of lurching in and out of ditches, the jeep finally lost a wheel. It was well below where we'd hoped to be driven and there was no choice but to continue on foot. We hope for a repair by tomorrow... As soon as we left, a light rain began and we quickly discovered how ill-equipped we were for the climb. Over a bleak lunch, we questioned ourselves for taking on this crazy journey and accused each other of leading a wild goose chase. We set off again in wet shoes—walking being the only way to stay warm as the altitude brought cooler air.

Soon, an ethereal fog met us and despair turned to delight. It was so thick that it was almost opaque after a few meters, so even the ground at our feet was shrouded in mist. It tasted sweet and felt as though we were breathing underwater. After hours of climbing, we almost forgot our goal. We wondered whether we would even be able to see the crater once we got to the top, or if we might fall right into it—

We read that Tambora can be translated from the language spoken in nearby Bima as "an invitation to disappear." It seems like a beautiful name, but it might actually be an insult directed at the people that once lived here, like "get lost!" We've been thinking about it all day, playing around at evaporating into the clouds.

Wish you were here...

Love, DH & JC

*A Year Without a Winter: An Invitation to Disappear* Dehlia Hannah, Julian Charrière, 2018

3 Progress

Dears,

Tambora's slopes are covered with a forest of flowering plants a meter or so high that our guides call Indonesian edelweiss. But edelweiss is becoming endangered by its own symbolic power. In Malay, it's called *bunga abadi* or *eternal flower*, and the tradition is that you give the flower to your beloved in order to wish them eternal life. We've been warned repeatedly not to pick it, but we've started eyeing it mischievously... thinking of you...

By running local newspaper articles through Google translate, we managed to read that a tourist was arrested recently for picking edelweiss here by the bushel, to sell in the market. It's a high-altitude plant that's already gone from many of Indonesia's volcanos. He was sentenced to two weeks of picking a proportional quantity of invasive species—quite a just and fitting punishment! If only we could be assured of the same, we joked, it would make a fine piece of performance art if we got caught and had to do the same. This little reverie about durational weeding got us thinking about the conflicting implications of the nascent tourism industry here—of which we partake, and to which we are contributing. For all the jet fuel, plastic water bottles, trash fires, and questionable labor conditions that it took just to get us here, what would be an elegant penance? We met in Antarctica, aboard the Antarctic Biennale. There, it is obvious that every step leaves a trace, and the onus is on the artist to make that trace worthwhile.

Wandering, with a mischievous spirit, we were at a loss, in certain moments, for a clear sense of purpose in this endeavor, whose difficulty and absurdity rises in tandem. Are we here out of a blind ambition whose true raison d'être will only disclose itself in time—or are we just lost?

Such moments attend all creative pursuits—As Captain Walton recognizes in the beginning of *Frankenstein*, what one needs in such moments is the company of true friends. For yours, we are eternally grateful!

Big Hugs, DH & JC

4 Anaphalis Javanica

*A Year Without a Winter: An Invitation to Disappear* Dehlia Hannah, Julian Charrière, 2018

PUNCAK KALDERA

5 Gunung Tambora, Puncak Kaldera

*A Year Without a Winter: An Invitation to Disappear* Dehlia Hannah, Julian Charrière, 2018

Dears,

The summit appeared suddenly and nothing could be as astonishing as the wall of mist that swept up along the edge, where heat from the caldera met the moist air. To the chagrin of our guides, we danced along the rim in a game of hide and seek, disappearing in and out of the fog. Under its blanket, the light was soft and diffuse—the sun a dimly glowing spot above. Just a few steps out, and the sun was blazing, drying our damp clothes immediately. Suddenly everything was sharp and bright, the colors super-saturated.

We've been reading so much about this volcano and staring down into its crater through satellite eyes—we know Tambora as a character in a global drama. Even here, everyone tells the story of Mary Shelley and the Year Without a Summer. It's a narrative that spread around the world as thick as Tambora's cloud did 200 years ago.

Looking out over the six-kilometer-wide caldera, we tried to imagine the tall mountain peak that used to be here, blown straight up into the stratosphere. So, this is how much rock and ash it takes to plunge the whole planet into darkness and chaos! Gusts of white smoke escape from the ground near a mineral-green lake below, the smell of sulphur a reminder that the volcano is still active. Our original plan was to hike down and camp at the bottom of the crater tonight, but we can't leave the fog—we want to see it at sunrise. We descended just a bit to explore before dark, partly climbing, partly riding the little landslides we caused on the loose dirt walls, sending debris tumbling down below us. A particularly beautiful lump of igneous rock flecked with molten metallic blue became our first souvenir.

Julian once made a piece about hydrocarbon emissions from the Canadian tar sands called *A Sky Taste of Rock*—which is exactly what Tambora gave the world. Having traced the taste of this rock back to its source, we are hungry to touch its reality. It's a hunger that's always there, but usually we satisfy it partially, with pictures and statistics—the fast food of experience.

This trip isn't long enough—you must come back with us!!!

Love,    JC & DH

6 Updraft

*A Year Without a Winter: An Invitation to Disappear* Dehlia Hannah, Julian Charrière, 2018

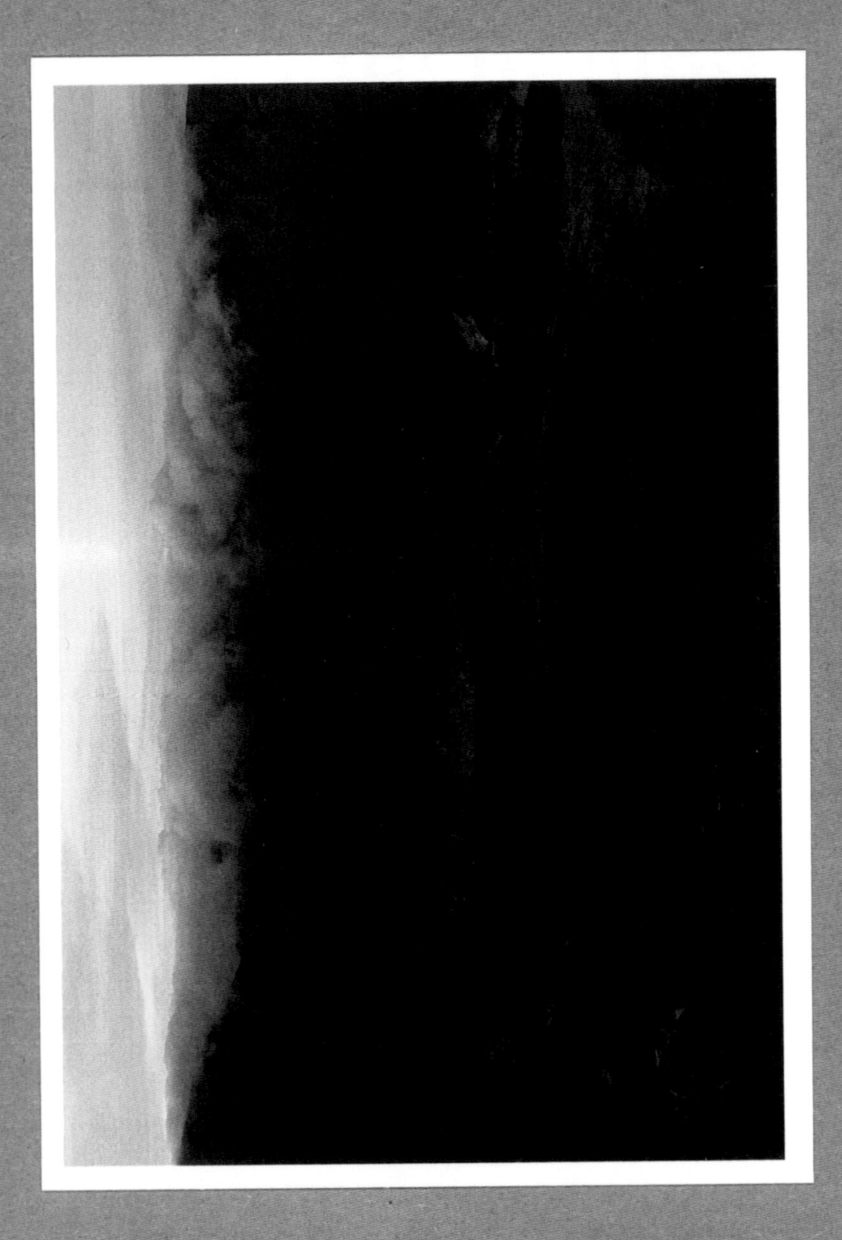

Dear Julian,

As the sun set and clouds gathered in a crescent around the edge of the crater, rolling down into the caldera on the sinking cool air, a memory flooded me. Since childhood I have harbored a confused vision, one certainly inspired by my father reminiscing about a summer he once spent in the Alps: I thought that if you climbed all the way to the top of a high mountain, you could step right off it, onto the clouds around the peak.

As we set off this morning, I faltered, suddenly drawing a blank where I wanted to share with you a sense of purpose. Instead, a flash of panic shot through me like a lightning bolt, and I was grateful for your reassurance. Settling into the cab of the jeep, I allowed the jostling of my body to take over my attention, and my thoughts privately turned towards a deeper feeling of emptiness.

As you know, I lost my father two weeks ago. Perhaps I should have cancelled this journey, but every fiber of my being refused. The last time we spoke, one of the few times in the last years of a difficult relationship, I told him about this trip. It delighted him, as I knew it would. It was

*A Year Without a Winter: An Invitation to Disappear* Dehlia Hannah, Julian Charrière, 2018

something he would do. He failed me in many ways—but he encouraged my wildest pursuits. After a frank discussion with my family, we agreed to postpone his memorial until we return.

By the time the fog met us on our ascent, my thoughts were elsewhere—fully absorbed by the comedy of our weather and gear tribulations. It is only now that I realize that nowhere in the world could this misremembered story of my father, walking off the top of a Swiss mountain onto a cloud, come closer to true than here on top of Tambora, today. Only the most outrageous promises my father made were the ones that turned out to be true.

All along we've been seeing signs that we can't read, like the little bits of plastic tied onto the edelweiss to mark a path. All the coincidences that brought us here are starting to read like omens. I wonder if this was a special day for the mountain too...Thanks so much for making this trip with me.

Love, Dehlia

7 Downdraft

Dears,

Last night, we zipped ourselves into our tent with only our heads sticking out and watched unfamiliar stars come out across the vast expanse of clear dark sky. We were awoken before sunrise by shouts and were surprised to discover that we had company, up on the mountaintop. In the distance, we spotted a red and white flag flapping on a pole carried by a few men climbing towards the highest point on the ridge.

Stiff and sleepless, having chosen the view over shelter from the wind, we gulped our instant coffee and packed in a hurry to start our return trek. The mountain looks so different in bright sunlight! Colors appeared on the loose ground—shades of gray, purple, brown, and yellow that we didn't notice before. Thinking Julian might make use of it later, we shoveled handfuls of dirt into empty bottles and stuffed our pockets with rocks until our pants were falling down. It was hot and the sun burned, as we originally expected. We couldn't wait to get below the clouds again.

Arriving back to the spot we were supposed to drive to the first day, we were surprised to find the jeep fixed and a group of twenty or so teenage boys, and one girl, accompanied by a smartly dressed police officer—all of whom insisted on posing for pictures with us. The policeman, who spoke English well, told us that today is Independence Day. On August 17, 1945, the Proclamation of Indonesian Independence from the Netherlands was issued. They were climbing the mountain to celebrate.

Volcanoes host all kinds of mythologies—angry gods, departed spirits, portals to new worlds, dragons born in flames...In the modern era, they're readily cast as figures of resistance to colonial orders and logics of globalization whose infrastructure they disrupt, sending up fire, dust, and smoke signals to the whole world.

One more lost wheel, down the mountain, seven-hours across the island, and we're filthy and exhausted beyond words, but excited to catch a ferry early tomorrow. It will take another eight hours to get there but we have to see the Komodo dragons! After all, this project is all about chasing monsters...

Bisous! See you soon! JC & DH

*A Year Without a Winter: An Invitation to Disappear* Dehlia Hannah, Julian Charrière, 2018

9 Varanus komodoensis

*A Year Without a Winter: An Invitation to Disappear* Dehlia Hannah, Julian Charrière, 2018

บจ 7315

*A Year Without a Winter: An Invitation to Disappear* Dehlia Hannah, Julian Charrière, 2018

Dears,

First of all, we must tell you that the ferry was named *Safety First*—a familiar mantra from Antarctica. After watching a dive boat sink in Thailand some weeks ago, we hoped it would live up to its name! At Komodo National Park you must be accompanied by a ranger—to protect us from the dragons and nature from us. We chose the long hike (compared to the last few days, a gentle stroll), constantly on the lookout for the fabled human-sized lizards. We found them, crowded lazily in the shade of the park canteen. They're not dangerous, the ranger said, as we waited our turn to take pictures. They're well fed on kitchen scraps. Not that we blame them—

All over the region, we've been passing oil palm plantations and trucks hauling their big red fruit bunches. Massive forest fires burn here during the summer—they're set to clear peatland and jungle to make way for more of this cash crop, sometimes blanketing the whole region in choking smoke. The fires emit huge quantities of CO2 and destroy carbon sinks, a large portion of Indonesia and Malaysia's emissions. Instead of cooling the world, two centuries later a cloud of smoke comes from the same landscape to heat it up.

All week we've been breaking up our diet of *nasi goreng* with a nasty little candy bar that we bought by the dozen. The ingredients list is in Indonesian, but it dawns on us that the only reason its edible at all is probably because it's made with palm oil. And even if we could read the label, palm oil is often anonymous...it's everywhere, in everything—an environmental catastrophe in a snack.

On the boat ride, we talked and philosophized, and planned a series of Instagram posts to break the Internet—#aninvitationtodisappear. It's been nice to be offline for almost a week, and we're laughing at ourselves for our obsession with mediating our experiences, performing our lives for the camera so much that it's hard to take anything in. Sometimes you have to go far away in order to be present—and what does it mean, even, to be present in such a globalized, interconnected world?

Julian talked about his dream of starting an art movement called Expeditionism. You, fellow 21st-century wanderers, are part of it. It is time to write the manifesto!

Love, DH & JC

10 Vegetable Fat

Dears,

We're becoming horrified and fascinated by the palm plantations. In a way, we overlooked them on our way to Tambora—not that we didn't see them, but we were focused on getting to the volcano, the national park, the beach. Driving past the endless rows of trees flash by like a flip book, a perfect grid stretching in every direction. The play of sunlight under the tree canopy is deeply ataractic. Filtered through the fronds and the humid air, a dark green light penetrates to the forest floor, giving the whole place a cinematic feeling. There's something haunting about these plantations, as though there are ghosts from the colonial era floating around...and vampires, from the global market, sucking the oil into snacks, cosmetics, even biodiesel engines. And yet, we're so used to seeing palm trees as signs of tropical paradise that it's hard not to be seduced by it all...

We know everything that's wrong with this industry, yet we're here, watching people make their living off it—an alternative to working in

*A Year Without a Winter: An Invitation to Disappear* Dehlia Hannah, Julian Charrière, 2018

the tourism industry. These are the *other* palm trees! We've been talking about how to activate this space, trying to imagine how an artwork might crystalize the geopolitical contradictions embedded in it. There's an uncanny emptiness about these monoculture plantations...they appear almost automated...and all so that we can eat the same packaged food all over the world. Even as we come looking for exotic sites, the homogenization of taste is reflected in the landscape. Our consumer desires are here, like the volcano was there, halfway around the world, long before anyone knew that it had saturated the atmosphere with dust.

On Tambora, we played with becoming invisible in the fog and using our voices to find each other. Now we're thinking about how to capture our absent presence amidst the oil palms...

Come with us!

Into the woods!

Love, DH & JC

11 Elaeis Guineensis

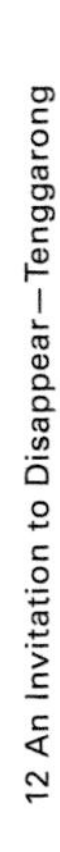


*A Year Without a Winter: An Invitation to Disappear* Dehlia Hannah, Julian Charrière, 2018

# Acknowledgments

I extend the warmest thanks to the contributors to this volume, who have sustained the collective thought experiment of *A Year Without a Winter* over the past three years and stretched my imagination in unforeseen ways. I am especially grateful to Cynthia Selin for her shared enthusiasm in initiating this project and for introducing me to the methodologies of scenario planning and futures research. Over the course of its evolution this project has benefited from the support of many collaborators and interlocutors: David Guston, head of Arizona State University's School for the Future of Innovation in Society and Frankenstein Bicentennial project; Sha Xin Wei, whose invitation to lead the *Atmosphere and Place* research network from ASU's School of Arts, Media, and Engineering occasioned initial research towards this book; Melissa Bukovsky and Linda Mearns of the National Center for Atmospheric Research; Angela Pereira of the European Commission-Joint Research Centre, for hosting the *Fictions and Policy* workshop; Jacob Lillemose for joining me at Arcosanti and Biosphere 2 for a preparatory rehearsal dare and hosting an exhibition of disaster fashion at his Copenhagen gallery *X and Beyond*; Ed Finn and ASU's Center for Science and the Imagination for sponsoring the reeneactment of the "dare"; to Ariel Anbar, Paul Hirt and Mark Lussier, Emily Mendelsohn, and Matt Bell for their participation; and especially to Joey Eschrich and Brenda Cooper for commissioning and editing an extraordinary collection of stories. At ASU, this project was supported by US National Science Foundation cooperative agreement #0937591. Thanks to Felix Reid for hosting me for the second half of this project as a guest researcher in the stimulating environment of Aarhus University's Center for Environmental Humanities and Laboratory for Past Disaster Research. Thanks to James Graham, Isabelle Kirkham-Lewitt, and Columbia Books on Architecture and the City for their unflagging support for this project, and Lauren Francescone for materializing the ideas graphically. My deepest thank to James for helping me to envision the implications of this book and allowing it to morph into the monster before us today. On a personal note, I could not have conceived this project without the provocations of my Doktormutter, Lydia Goehr, of Columbia University's Philosophy Department, who taught me to seek out the dialectical tension in the foundational concepts of every field—in this case, climate. Richard Grusin of the University of Wisconsin-Milwaukee's Center for 21st Century Studies offered indispensable support for early research on the aesthetics of climate change at a moment in which the topic appeared crass. I would also like to thank Alexander Ponomarev, founder and commission of the Antarctic Biennale, my participation in which was a transformative moment in the project. I have been inspired by many artists, friends, colleagues, and strangers, with whom I have discussed the stories of *Frankenstein* and Tambora at academic conferences and in taxis, elevators, and clubs in the wee hours of night. Thanks especially to Karolina Sobecka for her audacious artistic engagement with geoengineering; and to Julian Charrière for joining me on an expedition to Tambora from which a rich and ongoing collaboration was born. In Nadim Samman I found a friend who transcends reality and fiction, and allowed me finally to take leave of any distinction between the two. Finally, special thanks in memorium to my father Jed Harris, whom I searched for on many mountains, and my mother Bette Druck, who always encouraged me to live life as theater.

*For Animikii.*

**Dehlia Hannah** is a Mads Øvlisen Postdoctoral Fellow in Art and Natural Sciences at the Department of Chemistry and Biosciences at Aalborg University-Copenhagen and a guest researcher at the School of Earth and Space Exploration at Arizona State University. She holds a Ph.D. in Philosophy from Columbia University with specializations in aesthetics and the philosophy of scientific experimentation. She deploys her philosophical training to curate exhibitions and write about artworks that engage with scientific methods and materials and explore emerging environmental imaginaries.

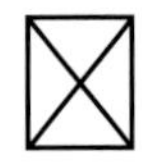

**Columbia Books on Architecture and the City**
An imprint of the Graduate School of Architecture, Planning, and Preservation

Columbia University
1172 Amsterdam Ave
407 Avery Hall
New York, NY 10027

arch.columbia.edu/books
Distributed by Columbia University Press
cup.columbia.edu

*A Year Without a Winter*
Edited by Dehlia Hannah
Fiction edited by Brenda Cooper, Joey Eschrich, and Cynthia Selin

Graphic Design
Lauren Francescone

Copyeditor
Ellen Tarlin

Proofreader
Hannah Manshel

978-1-941332-38-2



This book has been produced through the Office of the Dean, Amale Andraos, and the Office of Publications at Columbia University GSAPP.

Director of Publications
James Graham

Assistant Director of Publications
Isabelle Kirkham-Lewitt

Managing Editor
Jesse Connuck

Additional support for programming and editing has been provided by the Center for Science and the Imagination and the School for the Future of Innovation in Society at Arizona State University.

Library of Congress Cataloging-in-Publication Data
Names: Hannah, Dehlia, editor. Title: A year without a winter / edited by Dehlia Hannah. Description: New York : Columbia Books on Architecture and the City, 2018. | Includes bibliographical references. Identifiers: LCCN 2018024222 | ISBN 9781941332382 (pbk.) Subjects: LCSH: Creation (Literary, artistic, etc.) | Climate and civilization. Classification: LCC BH301.C84 Y45 2018 | DDC 304.2--dc23
LC record available at https://lccn.loc.gov/2018024222